Richard Wisen  only Professorship in the Public Understanding of Psychology, at the University of Hertfordshire. His research into a range of topics including luck, self-help, deception and persuasion has been published in the world's leading academic journals, while his psychology-based YouTube videos have received over 150 million views around the world. He is the author of several books that have been translated into over 30 languages, including *The Luck Factor*, *Quirkology*, *Rip it Up* and the international bestseller *59 Seconds*.

NIGHT SCHOOL

Richard Wiseman

NIGHT SCHOOL

Wake Up to the Power of Sleep

MACMILLAN

First published 2014 by Macmillan
an imprint of Pan Macmillan, a division of Macmillan Publishers Limited
Pan Macmillan, 20 New Wharf Road, London N1 9RR
Basingstoke and Oxford
Associated companies throughout the world
www.panmacmillan.com

ISBN 978-1-4472-4840-8 HB
ISBN 978-1-4472-5933-6 TPB

1 3 5 7 9 8 6 4 2

A CIP catalogue record for this book is available from the British Library.

Printed and bound by CPI Group (UK) Ltd, Croydon, CR0 4YY

Visit **www.panmacmillan.com** to read more about all our books
and to buy them. You will also find features, author interviews and
news of any author events, and you can sign up for e-newsletters
so that you're always first to hear about our new releases.

To Douglas and Cameron

Contents

Lesson 2

HOW TO BE HAPPY, HEALTHY, WEALTHY, AND WISE

Where we discover the remarkable power of sleep,
uncover an international zombie epidemic, and find out
why you need to head to your bed right now.

Assignment: What does your sleeping position reveal about you?

Lesson 3

THE SECRET OF SUPER-SLEEP

Where we find out the truth about 'short-sleep',
discover how to get the best night's sleep of your life,
and learn how to sleep like a baby.

Assignment: The mouth and nose test

Lesson 4

ON SLEEPWALKING AND NIGHT TERRORS

Where we go for a stroll with some sleepwalkers, find out if
you can commit a murder in your sleep, and discover the
deadly downside of snoring.

Assignment: Mid-term exam

Lesson 8

SWEET DREAMS

Where we find out how to control your dreams, banish
nightmares, and have a lucid dream.

259

Conclusion

TIME FOR BED

Where we engage in some myth-busting, reveal ten things that
every adult and child in the country should know about sleep
and dreaming, and set out to change the world.

291

Notes

302

Acknowledgements

322

Appendix

Donald and the Elephant at School

323

Introduction

WAKING UP

Where we come face-to-face with the devil,
discover the pressing need for Night School,
and set off into the darkness.

Something deeply strange happens every day of your life. You close your eyes, become oblivious to your surroundings, and repeatedly journey into a fantastical world. In this imaginary realm you might fly around, spend quality time with your favourite celebrity, defend the earth against a zombie apocalypse, or watch in horror as all your teeth fall out. Eventually, you regain consciousness, open your eyes, and carry on with your life as if nothing strange has happened. Perhaps most remarkable of all, this is not a brief experience. On average, you sleep for a third of each day, and a quarter of this time is spent dreaming.

Unfortunately, few people have any idea what happens during this part of their lives. *Night School* takes you on an in-depth tour into the science of sleep and dreaming, and presents practical techniques that you can use to get the most out of the night. During our time together we will discover what happens to your brain and body every night of your life, uncover the mysteries of the human sleep cycle, learn how to overcome nightmares, discover how to enjoy a great night's sleep, and find out how your dreams have the power to change your life.

I first became interested in the science of sleep a few years ago after I started to share my bedroom with the devil. Our clandestine meetings happened about once a week and always took the same form. Shortly after falling asleep, I would wake up in a cold sweat, stare across my room, and see Satan standing in front of my wardrobe. Sometimes he would start to move towards me, and other times he seemed content to keep his distance. Either way, it was a terrifying experience. After about a year of these strange happenings I was invited to take part in a public event on ways in which psychology can improve your life, and was delighted to discover that I would be sharing the stage with a highly regarded sleep expert, Dr Chris Idzikowski.

Chris is an amiable fellow who has had a long and colourful career investigating many aspects of sleep, including the best way to overcome jet lag and whether it's possible to carry out a murder when you are asleep. After the panel had finished, Chris and I went for a drink and I took the opportunity to tell him about my regular sightings of Satan by my wardrobe. After going through all the gory details I asked Chris whether I was having some kind of recurring nightmare. He asked a few simple questions. Did I ever scream out? How quickly did I get back to sleep? Did I suddenly sit up in bed? Chris then calmly explained that I was not experiencing a nightmare, but rather a very different phenomenon known as a 'night terror'. To the uninformed, these two experiences appear very similar. However, years of research into sleep and dreaming has revealed that they have very little in common. I walked away with a few top tips on avoiding night terrors (more about these later in the book), and I am delighted to report that I haven't seen Satan since.

Intrigued by the ease with which Chris exorcized my devilish tendencies, I started to explore the science of sleep and dreaming. Over time, my casual interest developed into a deep fascination, and I tracked down increasingly obscure academic papers in dusty journals and met up with cutting-edge sleep researchers.

I discovered that for the past sixty years or so, a small group of maverick investigators have devoted their lives to the night, often working long and unsociable hours to uncover the secrets of the sleeping mind. Never ones to shy away from controversy, these nocturnal scientists have carried out several strange experiments, spending months living in underground caves, staging secret studies with a legendary rock band, monitoring people as they attempted to set world records for staying awake, and bombarding entire villages with night-time messages. Inspired by this work, I carried out my own research, staging a mass participation experiment to discover whether people can take control of their dreams, assembling the world's largest dream bank (which now contains millions of reports), and creating the ultimate sleep environment.

For centuries, most people adopted a 'nothing to see, move on' approach to the night. They assumed that your sleeping mind is dormant, and that your time in bed has no real impact on your life. More recently, the scientific study of sleep and dreaming revealed that nothing could be further from the truth. In fact, each night you embark on an extraordinary journey that influences how you think, feel, and behave when you are awake. After years of tireless research, sleep scientists managed to map every stage of this fascinating journey, including which parts of your brain jump into action when you fall asleep, how to banish nightmares, and what your dreams really say about your psyche.

The work has, however, also uncovered the dark side of the night. Increased workloads, twenty-four-hour media, and permanent Internet access has combined to create a world that now never sleeps. The statistics are staggering, with surveys revealing that a third of both British and American adults do not get the sleep that they need, and that the vast majority of children arrive at school overtired. In 2010, British doctors issued more than fifteen million prescriptions for sleeping pills, and around one in ten adults now regularly take some form of sleep-related medication.

This epidemic of sleep deprivation is having a catastrophic impact on our lives. Around a quarter of drivers admit to falling asleep at the wheel, and fatigue is responsible for thousands of fatal road accidents each year. Poor sleeping habits also reduce productivity, prevent learning, disrupt relationships, cramp creative thinking, and sap self-control. As we will discover later in this book, some of the latest research suggests that poor sleep in adults is also associated with depression and obesity, and may cause children to exhibit many of the symptoms associated with attention deficit hyperactivity disorder (ADHD). Worst of all, even a small lack of sleep can have a detrimental effect on health, and is linked to an increased risk of heart disease, diabetes, high blood pressure, and an early death.

It doesn't have to be like this. As I continued my exploration into sleep science, I realized that much of the research could be used to create techniques to help those struggling with the night. Also, in the same way that these techniques can help move people from being a poor to good sleeper, so they can also help others go from good to great. During my research I uncovered the existence of super-sleepers. These people are able to fall asleep whenever they want, wake up feeling refreshed, and have lots of sweet dreams. Compared to most, they are significantly more likely to be happy, healthy, and wealthy. I believe that almost everyone can improve their sleep and make the most of their dreams and, in doing so, become a super-sleeper.

For years the self-development movement has focused on improving people's waking lives. *Night School* reveals how everyone can make the most of the remaining third of their day. It's time to reclaim the night, to change your life while you are sound asleep, and to wake up to the new science of sleep and dreaming. Welcome to Night School.

ASSIGNMENT

It's all in the timing

Throughout *Night School* you will be invited to carry out a series of specially designed questionnaires and exercises. Some are designed to be fun, and others have a more serious side.

This first one will take just a few moments, and involves completing the following questionnaire.[1] Don't spend too long thinking about each question, but instead just circle the response that instantly feels right. Oh, and please ignore the numbers under each of the possible answers (shown in italics); we will come back to those later. OK, away you go.

1) If you were free to plan your evening, and had no commitments the next day, what time would you choose to go to bed?

Before 21.00	21.00–22.30	22.30–00.00	00.00–01.30	After 1.30
1	*2*	*3*	*4*	*5*

2) If you were free to plan your day, what time would choose to get up?

Before 06.30	06.30–08.00	08.00–09.30	09.300–11.00	After 11.00
1	*2*	*3*	*4*	*5*

---

3) In general, do you find it easy to get up in the morning?

Definitely yes	Yes	Uncertain	No	Definitely no
1	2	3	4	5

4) Imagine that you have to do two hours of physically hard work. If you were entirely free to plan your day, in which of the following periods would you choose to do the work?

08.00–11.00	11.00–13.00	13.00–15.00	15.00–17.00	17.00 – 19.00
1	2	3	4	5

Lesson 1

INTO THE NIGHT

Where we find out what happens to your brain and body
every night of your life, discover how to overcome jet lag,
and learn the 'ninety-minute rule'.

Welcome to your first day at Night School. In this lesson we will explore two ideas that underpin the whole of sleep science and, in doing so, find out what happens to your brain and body every night of your life. We begin by examining the electricity coursing through your brain right now, then we'll meet an eccentric German professor who spent his life attempting to prove the existence of telepathy, and finally we'll spend the night in a modern-day sleep laboratory.

The personification of static

I would like to start by telling you something that has been playing on my mind for quite some time. You are amazing. There, I've said it and there's no going back. However, before you start to feel overly smug, there's something else that I need to say. I think your closest friend is amazing too. In fact, I think that everyone you know is remarkable (except for John in Accounts, who is actually quite annoying). And why are you all so jaw-droppingly wonderful? Because each of you owns one of the most wondrous, and complex, objects in the universe. This object has cured disease, put men on the moon, and created breathtaking works of art.

It allows you to see the world and listen to music, to make momentous decisions and move around, to laugh and to love. This remarkable object is sitting between your ears at the moment, quietly whirring away and allowing you to read this sentence. I am, of course, referring to your brain. (If you hadn't figured that out by now, I retract my initial compliment.)

Although everyone has a brain, most people are unaware that their mind runs on electricity.

If you were to slice off the top of someone's skull you would come face-to-face with what appears to be a large lump of pink jelly. Study any section of this strange substance under a high-powered microscope and you will find that it's made up of lots of tiny cells called 'neurons' (see the diagram of a neuron below). Each neuron consists of three main sections – (1) 'dendrites', finger-like fibres that receive signals from other cells; (2), 'axons', which pass signals to other cells; and (3) a 'cell body', which controls everything. Together, these deceptively simple cells are responsible for every thought that has crossed your mind and every emotion that you have experienced.

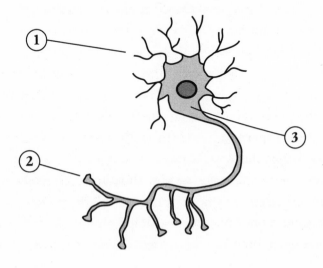

Neurons are little electronic messaging systems. When the dendrites receive a signal from a neighbouring neuron, the cell body springs into action and sends a tiny electronic pulse down its axon and on to the surrounding cells. These electronic messages are zipping around your head at this very moment, sometimes at speeds in excess of 200 miles per hour. Neuroscientists now believe that there are about 20 billion neurons in the average brain, and more than 160 trillion connections between them. Although any one neuron only creates a tiny amount of electricity, their combined output is considerable, with the average brain generating enough energy to power a 20-watt light bulb.

Around the turn of the last century, scientists were aware that the brain ran on electricity, but couldn't figure out a way of measuring the tiny signals produced by groups of neurons. Enter the most curious of men, Dr Hans Berger.[1]

Born in Germany in 1873, Berger's life changed forever when he had a close encounter with a cannon. Berger had enlisted for the cavalry service in his twenties. During training, he was thrown from his not-so-trusty steed and landed in the path of a horse-drawn cannon. The driver of the artillery battery carried out a textbook emergency stop which left Berger badly shaken but unhurt. At the precise moment of the accident, Berger's sister had had a strange feeling that her brother was in danger and sent a telegram asking if he was OK. This was the only telegram that Berger had ever received from his family, and he struggled to write off the experience as a coincidence. Instead, Berger became convinced that the spooky event was proof of telepathy, and devoted his life to discovering how thoughts can travel from one mind to another.

Working alone, Berger was desperate to develop what he referred to as a 'brain mirror' – a system of sensors that could be placed on the scalp and used to measure the tiny amounts of electricity being generated by the neurons inside the skull. Berger's experiments were as time-consuming as they were frustrating, but he locked himself away in his laboratory and persevered in the face of failure (Diary entry, 1910: 'Eight years! Trying always, time and again.'). The German professor grew increasingly distant from his colleagues and came to be seen as a deranged madman. To devote as much time to his research as possible, Berger ensured that his life was highly automated and predictable, with one of his colleagues later noting Berger 'never overlooked a deviation from established routine . . . His days resembled one another like two drops of water. Year after year he delivered the same lectures. He was the personification of static.'

After two decades of disappointment, Berger made a series of technological breakthroughs that hinted at success (Diary entry, 1924: 'Is it possible that I might fulfill the plan I have cherished for over twenty years?'). After spending several more years refining his invention, Berger finally announced that he was able to reliably record brainwaves, and demonstrated the world's first fully functioning electroencephalogram (or 'EEG machine' for short).

Unfortunately, the academic community adopted a somewhat closed-minded response to Berger's invention. Convinced that it was impossible to detect such tiny amounts of electrical activity from sensors placed on the scalp, many of Berger's colleagues assumed that his findings were due to either error or fraud. After retiring from academia in 1938, Berger's health quickly deteriorated and he became deeply depressed. The maverick measurer of minds eventually took his own life in 1941, hanging himself in hospital.

Berger didn't ever prove the existence of telepathy. Instead, he left a far more wondrous and tangible legacy. Academics across the world eventually realized that he had made a genuine break-through, and began to take a closer look at his remarkable invention. One of the first in the queue was a Wall Street tycoon and eccentric researcher named Alfred Lee Loomis.

The palace of science

Born in 1887, Loomis was both an amazingly successful invest-ment banker and the last of the great amateur scientists.[2] As a child, Loomis was fascinated by puzzles, chess, and conjuring. As a young man he developed a passion for science, and eventually struck up a close working relationship with a well-known experi-mental physicist from Johns Hopkins University named Robert Wood. It was an odd but productive collaboration. At one point, for example, Wood built a large 'spectrograph' (a device designed to disperse radiation into a spectrum) in his barn, but discovered that the instrument's forty-foot tube was frequently ineffective because it became filled with spider webs. Wood and Loomis eventually came up with a strange but highly effective solution to the problem. Whenever the spectrograph became blocked, the intrepid duo would drop a cat in one end of the tube and place some food at the other end. As the cat made its way towards the food, its fur acted like a huge duster and removed the cobwebs.[3]

Loomis enjoyed his time in the barn and eventually decided to build his own private research institute. In the 1920s, he bought a large mansion in New York State, and set about cre-ating his 'palace of science'. Over the next decade, he fitted out his mansion with cutting-edge technology, and played host to some of the world's best-known scientists, including Niels Bohr,

Guglielmo Marconi, and Albert Einstein. Loomis made several important scientific and technological discoveries, including playing a key role in the development of radar, inventing a new way of meas-uring the muzzle velocities of guns, and helping to create ground-controlled approach systems for aircraft.

In the mid-1930s, Loomis heard about Hans Berger's remarkable invention and wondered whether it could be used to investigate sleep. He constructed his own EEG machine, and invited overnight visitors to his palace of science to have their brains monitored. Within a year, Loomis discovered that people's brains are not dormant when they are asleep but instead produce a small number of distinct types of waves. Additional work revealed that these waves occur in a highly predictable pattern throughout the night (we will discover more about this pattern later in this lesson). Although identifying these different stages of sleep was a remarkable step forwards, one final mystery remained. This last piece of the puzzle only fell into place twenty years later, and was the result of one of the most important experiments conducted in the twentieth century.

What's in a wave?

The brainwaves that are detected by EEG machines have two main features: amplitude and frequency. Both of these features are illustrated in the diagram opposite.

The amplitude (1) is simply the maximum amount of energy the wave has, and the frequency (2) is the number of times that the wave repeats each second. The frequency is usually measured in units referred to as hertz, or 'Hz' for

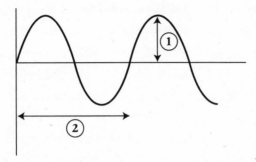

short. To understand the difference between amplitude and frequency, it's helpful to sing a little song. Please sing 'Laaaa' in a deep voice. You have just produced a low-frequency note that would look something like the top line in the following diagram.

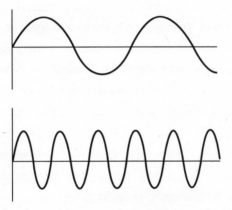

Now please sing 'Laaaa' again, but this time in a much higher pitch. If you were to plot the frequency of your voice now, you would get something that resembles the lower line in the diagram above.

Finally, try producing two more 'Laaaaa's, ensuring that they are the same pitch, but that one is much louder than the other. This time you are changing the amplitude of the wave, rather than the frequency, and so the quieter note would look like the top line in the diagram below, while the louder one would look like the bottom line.

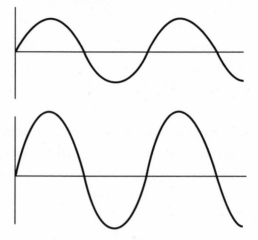

It's the same with brainwaves. Each brainwave can be classified according to the degree to which it is 'loud' or 'quiet', and 'low' or 'high'. In principle, this could result in millions of different kinds of waves. However, in practice, your sleeping and dreaming brain only produces a handful of different waves. For instance, when you are wide awake your brain creates 'beta waves'. There are about twelve to thirty of these waves produced each second, and so they show up on an EEG graph as rapidly changing squiggles.

When you relax, the frequency of these waves suddenly slows down until there are only about eight waves each second. The resulting waveform is referred to as an 'alpha wave' or 'Berger Wave' (named in honour of Hans Berger). During sleep, these waves become slower still, and we will take a closer look at each of the waveforms associated with the different stages of sleep later.

How to draw blood from a turnip

In 1951, Eugene Aserinsky was having a tough time of it.[4] Aged thirty, he was struggling to provide for his wife and son, with the entire family living in a small Chicago apartment heated by a single kerosene stove. Aserinsky had had an odd career path. After excelling at school he had skipped from college to college studying everything from Spanish to dentistry, failing to focus. He left education without a degree and found work as a high explosives handler in the army. After spending time taking his life in his hands, he decided to return to college. At the time, the University of Chicago had a reputation for accepting students with unusual backgrounds, and the unconventional Aserinsky was eventually enrolled in their physiology graduate programme.

When he arrived, he was less than delighted to discover that the only available academic advisor was an infamous and eccentric professor named Nathaniel Kleitman. The Russian-born Kleitman had dedicated his life to researching the science of sleep. In 1939 he had reviewed more than 1,000 scientific papers on the topic and written the then-bible of sleep research, *Sleep and Wakefulness as Alternating Phases in the Cycle of Existence*. Kleitman had also gained a considerable reputation for acting as his own guinea pig in the most challenging of conditions. In one series of studies, for instance, he had investigated how sunlight influenced sleep by spending a month in a large rock chamber deep within Kentucky's Mammoth Cave, by living on a submarine, and by exposing himself to almost continuous sunlight above the Arctic Circle. These endurance studies appear to have caused little long-term harm, with Kleitman dying at the ripe old age of 104 in 1999.

Kleitman met with Aserinsky and suggested that he study the

way in which babies blink when they fall asleep. Aserinsky found his endless observation of babies as 'exciting as warm milk' and, after spending months trying to 'draw blood from this research turnip', decided to call it a day. He switched his attention to brain activity and eye movements in sleeping adults. At the time most mainstream scientists thought that these eye movements were meaningless events that took place at random times throughout the night. Never one to follow the crowd, Aserinsky dragged an old EEG machine out from the departmental basement to his office and set to work.

He decided to start off by simultaneously measuring people's brain activity and eye movements throughout an entire night. This ambitious goal pushed the existing technology to its limit, as it required his dated EEG machine to operate smoothly for several hours. Aserinsky decided to carry out an initial test of the equipment by monitoring his son.

So it was that, on a cold night in December 1951, eight-year-old Armond Aserinsky found himself lying in a laboratory bed with his head covered in sensors. Some of the sensors measured his brain activity and others monitored the muscles around his eyes. All of this information was fed back to the EEG machine in a nearby room, where Aserinsky Sr sat watching several pens trace out the activity on a long roll of graph paper. This was no small-scale enterprise, with a single night of monitoring resulting in over half a mile of graph paper. As the night went on, Aserinsky Sr struggled to keep the out-of-date EEG machine up and running, unaware that he was just about to secure his place in history.

A few hours into the session Aserinsky Sr was surprised to see the pens suddenly start to scribble away, indicating that Armond's brain and eye muscles were highly active. Aserinsky Sr assumed that his son had woken up, and so went along to the corridor to see what had happened. When Aserinsky Sr opened the laboratory door he was amazed to discover that his son was sound asleep. Even more remarkably, this rather curious pattern was not a one-off affair, with Aserinsky Sr observing similar bursts of brain activity and eye movement throughout the night.

The next morning Aserinsky Sr tried to discover the cause of this mysterious activity. At first Aserinsky assumed that the old EEG machine was broken, and started the laborious task of checking the endless leads, dials, and valves. When he failed to find a problem he shared his results with Kleitman. Initially sceptical and perhaps suspecting fraud, Kleitman asked his student to re-run the procedure but this time using Kleitman's daughter as the participant. When the same pattern of data emerged, Aserinsky became more confident that he was on to something big, and labelled the curious phenomenon 'rapid eye movement' or 'REM' for short (Aserinsky originally thought about calling it 'jerky eye movement' but was worried about the negative connotations of the word 'jerk'). Intrigued, Aserinsky and Kleitman decided to find out what was going through people's sleeping minds when the EEG produced these strange nocturnal patterns.

Aserinsky arranged for a group of twenty volunteers to come to the laboratory. He woke them up whenever they entered REM state and interviewed them. Describing his findings in a now-classic paper ('Regularly Occurring Periods of Eye Motility, and Concomitant Phenomena, During Sleep'), Aserinsky noted that the vast majority of them reported a dream.

This paper had an enormous impact, with one leading scientist announcing that Aserinsky had discovered a new continent in the brain. For years the only dream reports available to researchers had come from people trying to recall their dreams each morning.

These reports were often patchy, incomplete, and unreliable. The discovery of REM changed the face of sleep science overnight and provided researchers with a direct route into the dreaming mind. As a result, scientists across the world started to investigate sleep and dreaming. Strangely, Aserinsky was not one of them. Ever the incurably curious polymath, Aserinsky switched to examining the effects of electrical currents on salmon, passing away in 1998 when his car swerved off the road and collided with a tree. Ironically, it is thought that he had probably fallen asleep at the wheel.

Aserinsky's remarkable discovery changed the world, and provided a pathway into the hitherto hidden world of dreaming. It was also the final piece of the sleep-science jigsaw and allowed researchers to map out exactly what happens to people every night of their life. To explore this map, it's time to visit a modern-day sleep centre.

Five amazing facts about dreaming

The discovery of REM has allowed sleep scientists to explore the mysteries of dreaming. Here are five of their strangest findings.

Dreaming in colour

The degree to which people experience colour in their dreams may depend on their childhood experience. Eva Murzyn, from the University of Dundee, asked people in their mid-fifties to rate both the amount of colour in their dreams, and how much black-and-white television they watched during their childhood. 25 per cent of those who only saw monochrome television when they were young dreamt in black and white, compared to just 7 per cent of those who had access to colour television.[6]

Up all night
Researchers have carefully measured the extent of male erections during dreaming and then compared this to the content of the dreams.[7] The findings show that erections happen during even the most mundane of dreams, and are not necessarily the sign of an erotic adventure.

Dreams of the blind
Research into the dreams of blind people has revealed that those who lose their sight before the age of seven experience dreams that contain almost no visual imagery, whereas those who become blind after they are seven years old have the same type of visually oriented dreams as sighted people. Also, those who are blind from birth report dreams that frequently involve vivid sensations of sounds, taste, smell, and touch.[8]

The importance of impotence
Nocturnal erections can help medics to determine the causes of impotence. If a patient does not get an erection during their sleep, then their impotence may be due to a physical problem that is best treated with drugs or surgery. However, if the patient has no problem 'staying up' all night then the problem is likely to be more in the mind.

You are blind when you dream
David Foulkes, from the University of Chicago, invited volunteers to his sleep laboratory, taped open their eyelids, and asked them to fall asleep.[9] When the volunteers started to dream, Foulkes tiptoed into the room and placed various

objects in front of their eyes, including an aluminium coffee pot and a card bearing the somewhat ironic message 'Do Not Disturb'. The volunteers were then woken up, asked to report their dream, and quizzed about what they thought had been happening right in front of their eyes. The volunteers saw nothing, and the objects didn't crop up in their dreams, suggesting that you become blind when you dream.

A touch of the Bram Stokers

Spend time in any psychology department and you will soon learn to spot the different kinds of researchers at work. The social psychologists are the ones who are unable to maintain eye contact, the memory researchers have forgotten where their offices are, and the persuasion experts will be arguing how best to split a bar tab. No matter how long you spend there, however, you are unlikely to spot that rarest of researchers, the sleep scientist.

This unusual breed enjoys a nocturnal and solitary existence. They arrive at their offices just as everyone else is heading home, climb into their beds when the rest of the world is waking, and often only meet one other person at work (and, all being well, that person will be asleep).

Stevie Williams is one of these researchers. Stevie is the head technician at one of the UK's best-known sleep clinics – the Edinburgh Sleep Centre. I first met Stevie a few years ago when we were both involved in a project examining whether psychics could dream about the future (they couldn't). Stevie is in his mid-thirties and, like most sleep researchers, has what I refer to as 'a touch of the Bram Stokers'. Although healthy looking, his skin has a thin, pallid quality, which I suspect is a direct result of his vampire-like existence.

Stevie had heard about my interest in sleep science, and kindly invited me to spend a night being monitored at his sleep clinic. Entering the clinic's sleep room is like being on the set of a stage play. On the face of it, everything looks like a normal bedroom or hotel room. Deep down, however, you have a strange feeling that all is not quite as it seems. Sneak a peek behind the bed and you will discover an endless array of sensors, tubes of gel, rubber caps, and miles of cabling. To the twenty-first-century sleep researcher this is exactly what you need to follow people as they journey into the night.

After I had changed into my pyjamas, Stevie glued about twenty small sensors in place on my scalp with a special gel, carefully connected a long wire to each sensor, and then gathered the wires together to form a strange-looking ponytail. The set-up looked bizarre but actually felt surprisingly comfortable. Stevie asked me to climb under the duvet, and then carefully placed the ponytail over the side of the bed. Finally, Stevie checked the positioning of the infrared camera that would record my every toss and turn throughout the night, and left the room.

The sleep centre's bed was remarkably comfortable, and after only a few seconds I found myself drifting off. The next moment Stevie was back by my bed gently waking me up. I assumed that it was the middle of the night and that something had gone wrong with the equipment. In fact, it was 7 a.m., and I had just enjoyed the best night's sleep that I had had for years. Stevie then asked me to change back into my civvies and meet him in his office.

I felt as if my mind had been in shut down-mode for the previous eight hours. No dreams. No activity. Zilch. Yet when Stevie showed me the EEG graphs from the night it was obvious that nothing could have been further from the truth. As he reviewed my data, it became obvious that my EEG trace was almost identical to the one produced by Aserinsky and Kleitman all those years ago. Modern-day sleep researchers refer to this pattern as the 'sleep cycle'. Over breakfast, Stevie kindly took me through each stage of the process.

It's just a stage he's going through

As we have seen, when you are wide awake your brain produces an erratic-looking EEG trace that contains between twelve and thirty waves each second. Soon after you climb into bed, the frequency of these rapidly changing squiggles slows down until there are only about eight to twelve waves each second. This type of trace is often associated with relaxation and meditation, and is known in the trade as 'alpha activity'.

After a few more minutes, your breathing will slow down, your eyes will roll from side to side, and the frequency of your brainwaves will become even lower. You are now entering **Stage 1** of sleep (see graph). You only enter this stage a handful of times

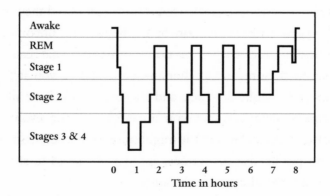

during the night, and each of these visits is very brief. During this stage your brain will be producing between three and seven waves every second or, to give them their technical name, 'theta' waves. If you are woken up during this stage, you are likely to feel like you weren't really asleep.

During your first encounter with Stage 1 sleep you might produce the odd twitch or two, and see illusory pinpricks of bright light or hear non-existent loud bangs (known as 'hypnagogic hallucinations'). Your muscles will also start to relax, and you are likely to experience a general 'loosening' of thought. Artists and writers have attempted to use this experience as a source of inspiration. For instance, surrealist Salvador Dali would lie down and place a glass on the floor. He would then put one end of a spoon on the edge of the glass and hold the other end between his fingers. As he drifted into Stage 1 of sleep, Dali's fingers would naturally relax and release the spoon. The sound of the spoon crashing into the glass would then wake him up, and Dali would sketch the odd images that were drifting through his mind.

This stage is also associated with a rather strange phenomenon known as the 'hypnagogic myoclonic twitch', which often starts with the sensation that you are falling before you suddenly find that your entire body has jolted itself awake. Around about 70 per cent of people experience these twitches, and they seem to be associated with exhaustion or sleeping in an uncomfortable position. Sleep scientists are not quite sure what causes the twitches, with some researchers arguing that as you fall asleep your muscles begin to relax, and the brain somehow misinterprets this as evidence that you are falling. Some evolutionary psychologists speculate that it might have developed from a time when people fell asleep in trees, and was designed to stop them falling out when they slept like a log.

You only experience Stage 1 of sleep for between two and five minutes. As you drift into the next stage of sleep, your heart rate slows down and your body temperature lowers. The 'theta waves' are joined by brief bursts of electrical activity known as 'spindles' and 'k complexes'. These appear to play a vital role in blocking out any external stimuli (such as a noise outside on the street) and internal stimuli (such as feeling a tad peckish) that might otherwise wake you up. You have now reached **Stage 2** of sleep. During this time almost all of your muscles, including those in your throat, start to relax, which can cause you to mumble or snore. Your brain is also taking a well-earned rest too, with a lowering of activity in areas associated with thought, reasoning, language, and problem solving. As we will discover in a later lesson, this stage is vital to learning physical activity, such as mastering a music instrument, new dance, or sporting skill.

Researchers often group the first two stages of sleep together, and refer to them as 'light sleep'.

After about twenty minutes in Stage 2 of sleep, your brain and body becomes especially relaxed, and you enter **Stage 3** and **Stage 4**. At this point your brain activity is at a minimum, resulting in very slow-moving 'delta waves' (only about one or two waves each second). Together, these stages are referred to as 'deep sleep' or 'slow-wave sleep'. During this time you will be almost completely cut off from the outside world (unless you happen to smell burning, someone says your name, or you hear a very loud noise). It's extremely difficult to wake up someone when they are in deep sleep, and if you do manage it they are likely to feel groggy and disoriented for several minutes.

Deep sleep stages are vital to your psychological and physical well-being because they are associated with the production of growth hormones that help repair damaged tissue. Without these stages you would wake up feeling tired and grumpy. These stages are also important for consolidating important information from the day, and are also associated with sleepwalking, sleep talking, and night terrors. When Stevie looked at my EEG trace he could see evidence of my tendency for night terrors. During deep sleep it's unusual for people to move around, but the recording from the infrared camera showed that I often moved my hands and arms.

Sleep scientists classify the first four stages of sleep as 'Non-REM' (or 'NREM') because they don't involve the type of rapid eye movements associated with dreaming. But does this mean that there is nothing going through your head during this time? If you are woken up from NREM sleep you are highly likely to report some kind of random, fragmented thought. This might take the form of a single word, or concept, and lack the strong sense of storyline that we commonly associate with dreaming.

After around thirty minutes in deep sleep, something very strange happens. Your brain and body move rapidly back through the different stages until you reach Stage 2. Then, instead of being relaxed, your heart starts to race, your breathing becomes shallow, and your eyes dart from side to side. Now you are experiencing rapid eye movement, or **REM**. During this time your brainstem completely blocks any bodily movement to prevent you acting out your dreams. If you were to be woken up now you would almost certainly describe a vivid dream. It is also quite likely that your sexual organs will be going into overdrive, with men gaining an erection and women showing increased blood flow to the vagina. Most people are in REM state, on and off, for about a quarter of the night, and this is sometimes referred to as 'paradoxical sleep' because the brain is almost as active as it is when you are awake. As we will find out later on in *Night School*, this stage plays a vital role in enhancing your memory, helping you deal with traumatic events, and seeing problems from a fresh perspective.

Having completed your first dream of the evening you move back down through the stages, and this NREM–REM–NREM sequence repeats itself again and again throughout the night. Each cycle takes around ninety minutes, resulting in an average of five dreams per night.

After each dream you might experience a very brief 'micro-awakening', wherein you are actually fully awake but for such a short time that you will not remember it in the morning. In a typical night, about 50 per cent of the time is spent in light sleep, 20 per cent in deep sleep, 25 per cent in REM, and 5 per cent having brief awakenings. The start of the night tends to be dominated by deep sleep, with relatively short dreams. However, as the night wears on the dreams become progressively longer and the periods of deep sleep correspondingly shorter. Indeed, in the second half of the night there is almost no deep sleep, and the REM can last up to forty minutes at a time.

The ninety-minute rule

Speak to sleep researchers and you will soon discover that most of them use a little-known trick to help them feel refreshed the next day. You will feel most refreshed when you awake at the end of a ninety-minute sleep cycle because you will be closest to your normal waking state. To maximize the chances of this happening, figure out when you want to wake up, then count back in ninety-minute blocks to find a time near to when you want to go to sleep.

> Let's imagine that you want to wake at 8 a.m., and wish to go to sleep around midnight. Chunking back in ninety-minute segments from 8 a.m. would look like this:
>
> 8 a.m. → 6.30 → 5.00 → 3.30 → 2.00 → 12.30 → 11 p.m.
>
> In this example, you should fall asleep around either 11 p.m. or 12.30 a.m. in order to feel especially refreshed in the morning.

After breakfast I thanked Stevie for talking me through the night, said goodbye, and walked out into the morning sun ready to face the day. Behind me, Stevie locked up the sleep laboratory and headed for his bed.

The sleep cycle plays a vital role in understanding what happens to your brain and body every night of your life. It is, however, only part of the picture. To fully appreciate the fundamental nature of sleep it's also important to get your head around a second key idea. It's time to discover what really makes you tick, and meet a man who changed the world by locking some plants in a cupboard.

The clockwork universe

The great eighteenth-century French astronomer Jean-Jacques d'Ortous de Mairan spent much of his career staring up at the sky.[10] In 1729, however, a rather more down-to-earth phenomenon caught de Mairan's attention. For centuries philosophers had observed plants opening their leaves during the daytime and closing them at night, and concluded that this curious behaviour was driven by sunlight. De Mairan wasn't convinced by their arguments, and decided to conduct a simple study that would put centuries of accepted wisdom to the test.

De Mairan decided to conduct his now-classic experiment with the help of the *Mimosa pudica*, a plant known for its rapid and highly predictable leaf movements. Each morning the *Mimosa pudica* opens and lifts its leaves, and every evening it closes and lowers them. De Mairan reasoned that if the *Mimosa pudica* were influenced by sunlight then it should stop moving when it is placed in total darkness. To find out if this was the case, he took one of the plants and locked it away in a pitch-black cupboard. Throughout the following few days de Mairan carefully lit a candle and peaked inside. Despite having no access to sunlight, the plant's leaves remained perky during the day and flaccid at night. His study had revealed that many of the world's greatest philosophers had made a terrible error, and that sunlight was not responsible for the *Mimosa pudica*'s behaviour.

At the time of his discovery de Mairan was working on several important astronomical projects, including exploring whether the colours of a rainbow were related to musical scales and trying to observe Venus's non-existent moons. As a result, the astronomer wasn't especially interested in publishing his work with the humble *Mimosa pudica*. In fact, the paper might not have seen the light of day, had it not been for his friend and fellow scientist, Jean Marchant. Marchant was convinced that de Mairan had made a major breakthrough, and insisted that the paper was published in the proceedings of the Royal Academy of Paris.[11] The article consisted of just 350 words. Nevertheless, it changed the science of sleep forever.

Over the next 200 years scientists carried out ever more complex versions of de Mairan's study in an attempt to discover the strange force controlling the opening and closing of plants.

After locking away thousands of plants in ever more secure cupboards, they ruled out every possible candidate, including temperature, humidity, and the earth's magnetic field. Eventually the researchers realized that plants were not responding to an outside force at all, but instead possessed a mysterious kind of internal clock that merrily ticked away regardless of what was happening in the world. Like a beautifully crafted timepiece, this internal clock worked on a twenty-four-hour cycle and ensured that the plant's leaves opened during the day and closed at night.

Flushed with success, the scientists then started to search for similar internal clocks hiding away in other forms of life. From the simplest single-celled organisms to the most amazing mammals, time and again they found what they were looking for. It soon seemed as if the whole of the natural world was controlled by clockwork. After decades of hard work the researchers finally reached the last item on their list: Homo sapiens.

Everyone has a natural tendency to wake up each morning and go to sleep every night, and researchers wondered whether this behaviour could also be the result of an internal clock ticking away in their brains and bodies. It was time for an experiment. De Mairan's initial groundbreaking study involved placing plants inside a pitch-black cupboard and then regularly observing their behaviour. Although carrying out the same study with humans might be fun (providing you weren't the ones in the cupboard), it wouldn't rule out other environmental factors that could influence the wake–sleep cycle, such as temperature, sound, and humidity. To stage de Mairan's experiment with humans, researchers needed to find a location that was completely isolated from the outside world and someone who was willing to stay there for a long time. Enter Michel Siffre, French scientist and adventurer extraordinaire.

Going underground

Born in 1939, Michel Siffre developed a passion for caving and science at an early age.[12] After graduating with a degree in speleology (the study of caves) in his early twenties, Siffre was on the hunt for an interesting research project. At the time a team of geologists had just discovered a subterranean glacier deep inside the French–Italian Alps, and Siffre realized that this was the perfect location for a groundbreaking experiment into the possible existence of the human internal clock.

In 1962, the twenty-three-year-old French adventurer descended nearly 400 feet below the earth's surface and lived in the cave for two months. Throughout the ordeal Siffre regularly telephoned his above-ground team to tell them when he had just woken up or was about to fall asleep. The experiment wasn't easy. Enduring below-freezing temperatures and very high humidity, Siffre suffered from hypothermia and frequently had to dodge large chunks of ice that fell around his tent. Yet Siffre's daily diary shows that he only lost the plot on one occasion. Tired, lonely, and clad only in a pair of black silk stockings, he decided to sing loudly while twisting the night away.

Siffre's suffering paid off, and the results revealed that humans do indeed have an internal clock ticking away inside them. In the same way that the plants in de Mairan's experiment regularly opened and closed their leaves despite being deprived of sunlight, Siffre continued to go to bed and wake up roughly every twenty-four hours. Over the next few years other sleep researchers investigated this mysterious internal clock by locking themselves and others in increasingly isolated underground locations. Their results have revealed an extraordinary insight into what really makes you tick.

The rhythm of life

Earlier in this lesson we discovered that your brain is made up of billions of neurons. Now let's take a somewhat bigger perspective and examine how large clusters of these neurons affect how you think and feel. If we removed your brain from your skull, and sliced it vertically in half, you would see something that resembled the following diagram (albeit without the labels).

Let's take a very quick look at some of the main parts of your brain. Towards the front there are the appropriately named 'frontal lobes' that are responsible for lots of things, including your level of self-control and ability to plan. In the middle there is the 'amygdala', which plays a key role in controlling your emotions, while at the back is the 'occipital lobe', which analyses the information from your eyes, and allows you to see the world. There, that didn't take long did it? Oh, hold on, I missed out two important parts.

First, take a look at the tiny black dot towards the middle left of the diagram (1). That is your 'suprachiasmatic nucleus'. This

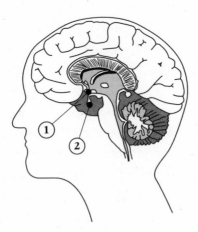

pinhead-sized group of about 10,000 neurons acts as your internal clock and merrily tick-tocks away every moment of your life.

Second, the other black dot towards the middle right of the diagram (2) is your 'pineal gland'. This pine-cone-shaped structure (thus the term 'pineal') is only about the size of a grain of rice but, nevertheless, has long fascinated philosophers, neuroscientists, and hippies. The famous seventeenth-century French philosopher René Descartes spent much of his life investigating the pineal gland, and eventually declared that it was the 'principal seat of the soul'. More recently New Agers have claimed that the area acts as a magical inner eye and is part of a holistic chakra system that promotes mystical awakening and ancient enlightenment (or something like that). But Descartes and the New Agers are wrong. In fact, the pineal gland is responsible for something far more important. At certain times of the day the suprachiasmatic nucleus causes the pineal gland to produce a sleep-inducing hormone called 'melatonin', making you feel drowsy and tired. Your internal clock sends these signals as part of a highly predictable pattern that repeats itself every twenty-four hours.* Sleep scientists refer to this pattern as a 'circadian rhythm', a term derived from the Latin word *circa*, meaning 'around', and *dies*, meaning 'a day'. Let's take a quick look at this pattern in more detail.

As with any cycle, you can begin to examine the effects of your internal clock at any point in the day. Let's start at 6 a.m.

———————

* In reality, the situation is slightly more complicated. In addition to the effect of the internal clock, your level of alertness is also determined by a second mechanism known as the 'homeostatic sleep drive'. In the same way that you feel hungry when you haven't eaten for a bit and satisfied once you have had a meal, so this drive builds up the longer you stay awake and quickly dissipates when you sleep. This drive interacts with your internal clock to produce the graph shown opposite.

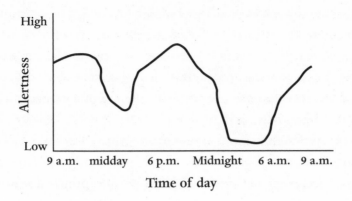

At this godforsaken hour most people feel pretty sleepy. However, for the next five hours your internal clock will make you feel steadily more alert, explaining why most people wake up between 7 a.m. and 9 a.m., and feel pretty good during the first few hours of the day. Starting at around about 11 a.m., you will slowly become less and less alert, leading to an all-time daytime low at about 3 p.m. If you are able to nap, this is the perfect time to nod off. However, do make the most of your nap time because this dip will only last for about an hour or so, and from about 4 p.m. onwards you start to become ever more energetic and perky, peaking at about 8 p.m. Then, from about 9 p.m. onwards, you experience a slow decline in energy levels, which encourages you to fall asleep just before midnight. Finally, these low levels continue throughout the night, and at 6 a.m. the following morning the entire process starts to repeat. And so the pattern continues throughout every day of your life. Like the gentle ebb and flow of the tide, your level of alertness gently rises and falls throughout the day.

Years of research have provided several important insights into this process.

There has been, for instance, work examining sleep and

babies. We are not born with Circadian Rhythms 1.0 pre-installed. Instead, newborns take lots of randomly timed naps throughout the night and day, resulting in them sleeping for about sixteen hours a day, and the rest of the time making as much noise as possible. As you might imagine, this is a joy to be around. However, the good news is that their circadian rhythms develop quite quickly, and within about six months they sleep more during the night-time.

Circadian rhythms undergo a major shift during our teenage years. Contrary to popular belief, teenagers that appear to be super-glued to their beds are not simply being lazy. During adolescence, circadian rhythms often shift by about three hours, resulting in teenagers not feeling sleepy until very late at night and not being able to prise themselves out of bed until late morning.

However, perhaps the most frequently researched aspect of the topic has examined the variation between the timing of people's circadian rhythms. Sleep scientists refer to this as your 'chronotype', with everyone falling on a continuum between those that naturally want to head to bed late at night and get up in the middle of the day (extreme 'owls'), and those that fall asleep early in the evening and spring out of bed at the crack of dawn (extreme 'larks'). Over the years researchers have developed various ways of measuring people's owl–lark tendencies, including monitoring their body temperature and levels of the naturally produced hormone melatonin in their blood. One of the most widely used ways of classifying people involves asking them to complete a questionnaire about their preferred bedtime, how they feel in the morning, and whether their heads rotate through 360 degrees. I asked you to complete one of these questionnaires just before the start of this lesson (page 8). Take a look at your answers now.

To score the questionnaire, add up the numbers (shown in italics) associated with each of your answers, and then use the following scale to discover where you fall on the lark–owl continuum.

4–6	7–10	11–13	14–17	18–20
Strong lark	Moderate lark	Neither owl nor lark	Moderate owl	Strong owl

Your chronotype is primarily determined by your genes, and so tends to run in families. It also has a large impact on the way that you think and behave. Perhaps not surprisingly, when it to comes to sleep, extreme larks like to be in bed by 10 p.m., wake up around 6 a.m., rarely need an alarm clock, and don't tend to nap during the day. In contrast, extreme owls like to go to sleep around 1 a.m., rise about 9 a.m., often set several alarms, and enjoy daytime napping.

In terms of when you do your best work and feel good, larks are most alert around noon and feel happiest between 9 a.m. and 4 p.m., whereas owls are most productive around 6 p.m. and feel at their best between 1 p.m. to 10 p.m.

There is a great deal of debate around the relationship between people's chronotype and their personality, but in general larks tend to be introverted, logical, and reliable, whereas owls are more extroverted, emotionally stable, hedonistic,[13] and creative.[14] On the downside, owls are also more likely to be untrustworthy, psychopathic, and narcissistic.[15] Perhaps not surprisingly, these differences have a dramatic effect on people's personal lives with, for example, research showing that, on average, owls have four times as many partners as larks during their lifetime. People's chronotype also affects their relationship with food, with larks showing a strong tendency for eating breakfast within thirty minutes of waking and owls being much more partial to a midnight feast. Unfortunately, lots of large evening meals tend to take their toll over time, causing more owls than larks to suffer from obesity.

There is also a strong relationship between chronotype and academic performance, with larks getting higher grades than owls. Researchers originally thought that this was due to larks being more intelligent than owls. In fact, the early start times adopted by most schools means that owls are often studying, and being examined, when they are not feeling at their best.[16] Because of this, several education experts think that there is a strong case for measuring pupils' chronotypes, and then using this information to help maximize their performance by, for example, scheduling lessons and exams at more appropriate times of the day.

School children and students are not the only ones to suffer because of their chronotype. Chronobiologist Till Roenneberg, from the Ludwig Maximilian University of Munich, has argued that adults' internal clocks are also frequently out of step with their surroundings.[17] During the week, many office workers are expected to be at their desks by 9 a.m. This is fine for larks, but tricky for owls. Struggling to go to bed earlier in the night, those with owl-like tendencies often only get a few hours' sleep before having to get up for work. As a result, they spend much of the week feeling overly tired, and so have to try to catch up on their lost sleep during weekends. For larks, it's the weekends that cause the problem. Many people's social lives involve staying up late on Friday and Saturday nights. Even if larks do manage to keep their eyes open at this time, they often struggle to lie in the following day and so end up getting very little sleep. Roenneberg has argued that both phenomena result in a kind of 'social jet lag' that is leaving large numbers of people constantly feeling tired.

However, perhaps the most important work into our internal clocks has examined how they can be used to help treat those struggling to sleep.

Rhythm and blues

Like many timepieces, your internal clock is not entirely accurate. In fact, Siffre's study showed that your internal clock is probably slightly on the slow side, taking just over twenty-four hours to complete a single cycle. Left unchecked, this small difference would cause you to slowly drift out of sync with the actual time, and after just a few weeks you would feel sleepy at the start of the day and wide awake when darkness fell. To help prevent this happening your internal clock is re-set each day by several factors, such as when you eat, move around and, most important of all, the amount of light entering your eyes.

The light swirling around you right now not only allows you to read these words (and, for that matter, these ones too), but also has a dramatic effect on your brain. This light is entering your eyes and causing your retinas to produce tiny electrical signals that are then stimulating your suprachiasmatic nucleus and pineal gland. When these brain structures receive this stimulation they stop producing the sleep-inducing hormone melatonin, and so you feel alert and awake. However, if you were to turn out all the lights, your retinas would cease to stimulate these bits of your brain, and the subsequent release of melatonin would make you feel drowsy and tired. For this reason neuroscientists often refer to melatonin as the 'Dracula hormone' because it only emerges in the dark.

Unfortunately, some people's clocks fail to respond to these light cues, causing them to suffer from what is referred to as a 'circadian timing disorder'. Over time, those suffering from the disorder slowly drift out of sync with actual time, eventually either going to bed very late at night ('delayed sleep phase disorder') or extremely early in the evening ('advanced sleep phase disorder'). Unlike most people with insomnia, sufferers tend to enjoy several hours of uninterrupted sleep. However, on the downside they find themselves getting up very late in the morning or towards the middle of the night, which can become problematic if their jobs or social lives require them to keep normal hours. Treatments often involve people sitting in front of commercially available 'light boxes' that expose them to massive amounts of illumination in attempt to re-set their internal clocks, with those suffering from delayed sleep phase disorder seeing the light between 7 a.m. and 9 a.m., and those experiencing advanced sleep phase disorder doing the same between 7 p.m. and 9 p.m.

Research into circadian rhythms has also discovered what to do when our internal clock fails to tell the right time. If you jet around the world, you cross one or more time zones, and your internal clock is unable to keep up with the change. As a result, you soon start to experience an annoying phenomenon known as 'circadian dysrhythmia' or, as most people refer to it, 'jet lag'.

Let's imagine, for example, that you are going to take a six-hour flight from London to New York, and that you will leave London at noon. When you arrive in New York your internal clock will think that it is 6 p.m., and so you might feel a tad tired and ready for an evening meal. Unfortunately, the actual time in America will be 1 p.m. and everyone there will be all alert and tucking into their lunch. As a result you will be suffering from an east-to-west lag that sleep researchers refer to as 'phase delay'. Now let's imagine that you had instead flown from New York to London, leaving New York at noon. In this scenario you would arrive in London thinking it is 6 p.m., and ready to head out on the town for a bite to eat. Unfortunately, the actual time in London will be 11 p.m. at night and so most people there will be heading to their beds. As a result, you will suffer from a west-to-east lag called 'phase advance'.

Over time, the light levels at your new destination will slowly re-set your internal clock to the correct time. However, before then, life can be tricky, with people feeling overly tired, thick headed, and ill. Phase delay is far less disruptive than phase advance, and so flying east to west usually creates far fewer problems than flying west to east, but even relatively small differences can have a surprisingly large effect on the brain and body.* In one set of intriguing studies, Lawrence Recht from the University of Massachusetts and his colleagues examined the performance of North American major league baseball teams.[18] When a team had to fly east to west before a game they won 44 per cent of the time. In contrast, when they had to travel west to east they were victorious just 37 per cent of the time.

Jet lag is unlikely to be a problem if the time difference between your point of departure and destination is less than three hours. Similarly, if you are only going to be away for a couple of days or less, it might be easiest to stay on your 'home time'. For all other long-haul trips, the difference between your internal clock and the actual time may well cause you to feel disoriented, light-headed, and listless. Adjusting to a new local time takes about half a day per time zone if you are flying east to west, and

---

* The human circadian rhythm is not exactly twenty-four hours, but rather about fifteen minutes longer. Because of this, your brain and body is used to a very mild form of phase delay, in which you 'lose' about a quarter of an hour each day. Some researchers believe that this daily experience explains why travellers find phase delay less disruptive than phase advance.

two-thirds of a day per time zone if you are flying west to east. However, worry not, because help is hand, with researchers developing several methods for beating jet lag (see below).

Top tips to help overcome jet lag

- Make good use of the days before you fly by starting to shift your body clock to the time at your destination. If you are flying to the east, get up slightly earlier. If you are flying to the west, get up slightly later.

- If possible, book flights that will minimize jet lag by following the simple adage, 'Fly east, fly early. Fly west, fly late.'

- If you need to sleep during the trip, try to avoid sitting on the sunny side of the plane. For flights in the northern hemisphere, the sun will tend to be on the left side of the plane when you fly west, and on the right side when you go east. Several websites offer more precise guidance for specific flights.

- As soon as you board the plane, adjust your watch to show the time at your destination, and try to fit into this new time schedule as soon as possible. If it is time to sleep, get your head down. If it is dinner time, eat something.

- Some people believe that melatonin supplements can help control your sleeping patterns and thus help you adjust to new time zones. Research suggests that daily doses of melatonin can help alleviate jet lag, and that

---

short-term usage seems to have few negative side effects. Consult your doctor before taking any medication.

- When you arrive at your destination, adjust your circadian rhythm by using the following simple rules of thumb:
  — If you have travelled east, avoid the morning sun and seek out natural light in the afternoon.
  — If you have travelled west, seek out light throughout the entire day.

- If you really cannot keep your eyes open during the day, take a quick nap, but set your alarm to make sure that it is no longer than two hours.

In this lesson we have explored two concepts that play a fundamental role in sleep. First, we discovered what happens to you every night of your life, and unlocked the mysteries of light sleep, deep sleep, and REM. In the second part of the lesson we turned our attention to circadian rhythms, examining how they determine where you sit on the lark–owl continuum, cause teenagers to struggle to climb out of bed in the morning, and can be used to overcome jet lag. Having dealt with these two fundamental concepts, in the next lesson we will turn our attention to what happens if you don't spend enough time in bed.

ASSIGNMENT

Assess your sleep

This questionnaire will take just a few moments, and involves selecting the answer option for each of the following questions that is most applicable to you.[19] Please don't spend too long thinking about each question, and ignore the numbers under each option.

1) Do you feel in control of your sleep? For example, can you fall asleep or stay awake whenever you want?

Definitely no	No	Uncertain	Yes	Definitely yes
1	2	3	4	5

2) During the day, do you often feel sleepy when you are, for example, driving, sitting in a meeting, or chatting?

Definitely no	No	Uncertain	Yes	Definitely yes
5	4	3	2	1

3) Do you tend to wake up during the night?

Definitely no	No	Uncertain	Yes	Definitely yes
5	4	3	2	1

---

4) When you do wake up in the middle of the night, do you tend to struggle to get back to sleep?

Definitely no	No	Uncertain	Yes	Definitely yes
5	4	3	2	1

5) Do you tend to have pleasant dreams?

Definitely no	No	Uncertain	Yes	Definitely yes
1	2	3	4	5

6) How would you rate the quality of your sleep?

Very poor	Poor	Uncertain	Good	Very good
1	2	3	4	5

7) How do you feel when you wake up in the morning?

Very groggy	Groggy	Uncertain	Refreshed	Very refreshed
1	2	3	4	5

Many thanks. More about this later.

Lesson 2

HOW TO BE HAPPY, HEALTHY, WEALTHY, AND WISE

Where we discover the remarkable power of sleep, uncover an international zombie epidemic, and find out why you need to head to your bed right now.

We are about to explore a remarkably simple idea that can help you feel happier, become more productive, lose weight, stop smoking, sharpen your thinking skills, be more creative, and boost your health. At times it is going to be a deeply strange journey, and along the way we will spend some time with the eccentric genius who lit up the world, find out what happens if you stay awake for several days, and discover how you may have just fallen asleep without realizing it. We begin by setting off to see the wizard, the wonderful wizard of Menlo Park.

Light-headed

In the previous lesson we discovered how light entering your eyes causes your brain to suppress the release of the sleep-inducing hormone called melatonin. As a result, you feel more energetic and lively. In contrast, darkness encourages your brain to unleash the hormone, which, in turn, makes you want to head for your bed. For millions of years the sun provided the only serious source of light on earth. For hundreds of thousands of years, sunlight entered your ancestors' eyes during the day and made them feel lively, while at night the darkness encouraged them to lie down and take it easy. This routine possibly evolved because hunting and gathering was especially efficient in the sunlight, and lying motionless in the dark was a good way of avoiding most potential predators. Every twenty-four hours the earth moved around the sun and, as sure as night follows day, your ancestors ran around in the light and fell asleep in the darkness. Then, about 150 years ago, everything suddenly changed.

Thomas Alva Edison was one of America's greatest inventors and businessmen. Born in 1847, he showed an early fascination for mechanics, chemistry, and hard work (hence his famous quote 'Genius is one percent inspiration, ninety-nine percent perspiration'). Although he received very little in the way of a formal education, Edison developed a flair for innovation at a very young age. When he was just twelve years old he set up a printing press in the baggage car of a train and started to publish the first (and perhaps only) newspaper to be produced on a train. Over the coming years Edison worked his way up the corporate ladder, and in 1876 opened his own laboratory in Menlo Park, New Jersey. Dubbed the 'Wizard of Menlo Park' by the popular press (and 'Mr Thomas Alva Edison' by the less popular press), Edison created a series of remarkable products that would define the modern world, including early versions of the microphone, telephone receiver, record player, movie camera, storage battery, and earmuffs.*

Edison was convinced that many Americans failed to fulfil their true potential because they spent far too much time sleep-

---

* I made up the last one. In fact, earmuffs were invented in 1873 by Chester Greenwood from Farmington, Maine. Chester created his first pair of earmuffs when he was just fifteen years old and later made his fortune selling 'Greenwood's Champion Ear Protectors'. Farmington, Maine, is now known as the Earmuff Capital of the World, and each December the town stages a parade in which local police cruisers are decorated to look like giant earmuffs.

ing. The great inventor frequently boasted that he only needed five hours of sleep a day, and once declared:

> We are always hearing people talk about 'loss of sleep' as a calamity. They better call it loss of time, vitality, and opportunities.

Driven by this love of activity, Edison decided to wage a war on sleep. During the early part of the nineteenth century, buildings and streets were lit by open fires, wax candles, oil lanterns, or gas lamps. These early forms of illumination yielded relatively small amounts of light, and so most people were forced to endure long periods of darkness after nightfall. Towards the latter part of the century, several scientists discovered that it was possible to create a much more powerful source of light by running an electric current through a metal filament. Unfortunately, the filaments that they used burnt out very quickly, and so proved completely impractical. Edison set out to develop a new form of filament that was cheap, effective, and reliable.

The Wizard of Menlo Park systematically experimented with a diverse range of materials, later recalling, 'Before I got through, I tested no fewer than 6,000 vegetable growths, and ransacked the world for the most suitable filament material.' After much trial and error, Edison placed a piece of carbonized bamboo in a glass vacuum bulb, ran an electric current through it, and created a light that burnt brightly for more than 1,000 hours.

Like moths to a flame, entrepreneurs across the land realized the potential worth of Edison's invention, and set about creating the complex infrastructure needed to bring the electric light bulb into homes and factories throughout America. In 1913, the first bulb to contain a coiled tungsten filament rolled out off the assembly lines and provided the type of economical, long-lasting, and bright illumination that we now take for granted. Just twenty years later, the vast majority of people in American and European cities had electricity coursing through their streets and homes. Midnight was no longer the darkest hour. Edison's brilliant invention had illuminated the world.

The light bulb had a huge impact on the fabric of society. At the simple flick of a switch, people across the globe were now able to see in the dark and so stay up long into the night. Millions of people made the most of these extra waking hours by socializing in their homes, or heading out to cinemas, theatres, restaurants, and bars. Artificial light also allowed manufacturers to open their factories around the clock, giving rise to shift work and the twenty-four-hour workforce. Night had become the new day.

The runaway success of the light bulb resulted in people being exposed to huge amounts of both natural sunlight and artificial light, resulting in the world remaining wide awake for ever-increasing amounts of time. In the last few years the situation has become far more extreme. Twenty-four-hour media has combined with increased workloads, social networking, smartphones and the Internet to transform us all into creatures of the night.

The statistics are staggering. In 2000, a large-scale survey examined the sleeping habits of more than a million people from 250 countries. Almost 50 per cent of respondents said that they required eight hours of sleep each night to feel well rested. Remarkably, only 15 per cent said that they actually managed this much sleep.[1]

In 1960, a survey of over 1 million Americans found that the majority of people were getting between eight and nine hours' sleep a night.[2] Around 2000, several surveys carried out by the National Sleep Foundation and others revealed that this figure had fallen to about seven hours.[3] In 2006, The Institute of Medicine estimated that around sixty million Americans had a chronic sleep disorder,[4] and recent surveys show that a third of Americans now get less than seven hours' sleep each night.[5]

The situation is no better in the UK, with a 2011 British survey revealing that more than 30 per cent of respondents suffered from insomnia or another serious sleep problem.[6]

We are now all living in a world that never sleeps. Because of this, scientists across the world started to examine what happens to people's brains and bodies when they become sleep deprived. Their results are enough to keep you up at night.

Shattered

One way of examining the role that sleep plays in your life would involve preventing you from ever sleeping, and then seeing what happens to your brain and body. This type of study has been conducted with animals, and the results have been both dramatic and deadly. In the 1980s, Allan Rechtschaffen from the University of Chicago conducted perhaps the best-known, and most disturbing, study into the topic. Rechtschaffen and his colleagues wired up a group of rats to a machine that measured their brain activity, and then placed each of the animals on a stationary disc above a bowl of water.[7] Every time the rat's brain activity indicated that the animal had fallen asleep, the disc would slowly rotate. This, in turn, forced the rat to wake up and move around in order to avoid falling in the water. Despite having access to more than enough food, within a week these sleep-deprived rats started to lose weight and their fur developed an unhealthy yellowish tinge. After a month, all of them had died, thus proving that sleep is essential for life.

Conducting a similar study with humans would be deemed highly unethical and, even if it were allowed, finding volunteers would probably prove a tad tricky (especially if they had heard about the results of Rechtschaffen's rat experiment). However, a handful of hardy souls have willingly endured slightly more modest amounts of sleep deprivation and, in doing so, revealed the vital role that sleep plays in your life. One of the first was a rather curious American disc jockey named Peter Tripp.[8]

Throughout the 1950s Tripp wowed radio listeners across America. Initially working out of Kansas City and billing himself as 'the bald kid in the third row' (a comment his mother made when she pointed out her newborn baby at the hospital), Tripp was an early and tireless supporter of the then-newfangled rock 'n' roll music. At a time when most disc jockeys prided themselves on being sedate and bland, Tripp shocked audiences with his edgy style and upbeat playlist. In 1955 he moved to New York, re-named himself 'the curly headed kid in the third row', and was soon offered the job of presenting one of the nation's best-known radio shows.

Then, in 1959, the thirty-two-year-old Tripp decided to garner some publicity by trying to stay awake for a remarkable eight days. If successful, he would create a record of his own by racking up the longest ever period of wakefulness. Tripp and his team set up a glass-walled portable radio studio in New York's Times Square and continued to broadcast his daily radio show throughout the ordeal. The publicity stunt was a huge success, with journalists from across the world reporting on Tripp's progress, and thousands of passers-by watching the sleepless disc jockey attempt to keep his eyes wide open. While the nation listened in disbelief to Tripp's broadcasts, a team of military medics kept the disc jockey under constant observation. These doctors took their task seriously, ensuring that they accompanied Tripp twenty-four hours a day, even when he visited the bathroom.

As the hours ticked away, it became harder and harder for Tripp to remain awake, and his behaviour became increasingly erratic. At one point, he invited his barber to give him a haircut. Despite knowing the man for several years, Tripp was extremely rude to the barber, eventually making him cry during the haircut (forcing Tripp to re-name himself 'the kid with badly cut curly hair in the third row'*).

After a few more days of deprivation, Tripp started to get trippy. He thought that there were mice running around the studio, that his shoes were jammed full of spiders, and that his desk had caught fire. At one point in the proceedings a local doctor came to examine him. Convinced that the medical man was actually an undertaker who had come to bury him, a half-naked Tripp ran screaming from the room. Towards the end of his ordeal, the exhausted radio host became uncertain about his real identity, and thought that he was simply impersonating a man called Peter Tripp. Many of these periods seem to correspond to times at which Tripp would have been dreaming had he been asleep, raising the intriguing possibility that he was actually having a dream despite being wide awake.

After staying awake for just over eight days, Tripp slept like a baby for twenty-four hours, and then seemed fine. However, Tripp's wife later said that her husband had become moody and depressed after the stunt. A few years later he became involved in a high-profile financial scandal, was fired from his job, and, in time, went through four divorces. To many it appeared as if Tripp

---

* I made up the bit about Tripp re-naming himself.

had achieved his world record only for the marathon of wakefulness to ruin his life.

Others have tried to survive with no sleep. In 1964, a seventeen-year-old American high-school student named Randy Gardner attempted to set a new world record by staying awake for eleven days. After a few days, the going got tough, and two of Gardner's friends agreed to help keep him awake by arranging trips to the local doughnut shop, playing loud music, and staging increasingly long pinball sessions. When Gardner visited the bathroom, his friends talked to him through the door to ensure that the record attempt wasn't going down the pan.

Commander John Ross, a doctor from the US Navy Medical Neuropsychiatric Research Unit, monitored Gardner throughout the record attempt and observed a range of psychological problems.[9] Towards the start of the stunt Gardner experienced many of the issues that had plagued Peter Tripp, including extreme moodiness, memory issues, and paranoia. On his fourth day of wakefulness he started to hallucinate, confusing a street sign for a person. In a later delusional episode, Gardner, who was white, became convinced that he was a famous black American football player. When friends tried to disabuse him of this notion, he accused them of making racist comments. Gardner was not the only one to suffer. After spending several nights awake, one of the scientists involved in the project became sleep deprived and was booked for driving his car the wrong way down a one-way street.[10]

By the end of the ordeal, Gardner's face appeared expressionless, his speech had become slurred, and he had developed a slight heart murmur. After setting a new world record he slept for around fifteen hours but, unlike Tripp, did not appear to experience any long-term changes to his personality. Ironically, in a recent interview Gardner admitted that he now occasionally struggles with insomnia.[11]

The curse of continuous consciousness

A rare genetic quirk has meant that over the course of the last few hundred years a handful of people have discovered the horrors associated with never being able to sleep.

In 1836, in a small town just outside of Venice, a middle-aged man named Giacomo mysteriously fell ill. Although he had been fit and healthy throughout his entire life, Giacomo was suddenly unable to sleep, began to suffer from dementia, and eventually died a few months later. Over the course of the next few centuries, many of Giacomo's descendants suffered exactly the same fate. For years, Italian doctors didn't know how to explain these premature deaths, and often attributed them to epilepsy or some kind of strange fever. Then, in the 1980s, researchers realized that these strange fatalities were actually due to an extremely rare genetic disease known as 'fatal familial insomnia' (or 'FFI').

The symptoms of FFI make for grim reading. The disease strikes without warning in middle age, and causes the sudden onset of complete insomnia. For months the sufferer is unable to sleep, and starts to experience severe panic attacks and hallucinations. They then rapidly lose weight, exhibit dementia-like loss of their memory, fall into a coma, and die. For much of the time sufferers are fully conscious, and so forced to endure the continuous agony of total sleeplessness.

Stanley Prusiner, a neurologist from the University of

California at San Francisco, spent much of his career investigating this strange illness. Prusiner eventually discovered that FFI is caused by an infection that makes the body destroy itself. This infection initially attacks the region of the brain responsible for regulating sleep, and so prevents the sufferer from ever losing consciousness. It then moves on to attack several other areas of the brain.

Researchers initially thought that Giacomo's descendants were the only people to have FFI, but have since found a handful of other families with the illness. The condition causes a great deal of suffering even before the infection starts, with many of the Giacomo's descendants struggling to find partners and unable to get life insurance. Although there is now a reliable test for the condition, there is no known cure.

As a psychologist, I have witnessed the effects of sleep deprivation first hand. A few years ago I was involved in a television show called *Shattered*.[12] The programme was part entertainment and part science, and involved ten contestants trying to stay awake for a week. The contestant who managed to avoid sleeping for the longest time stood to walk away with £100,000. To keep everything on a safe footing the producers enlisted a group of medical experts, and I was asked to help design various 'You Snooze You Lose' challenges. We came up with lots of fun ways of trying to get the contestants to fall asleep, including having them cuddle a giant teddy bear, listen to a bedtime story, watch paint dry, and sit through an incredibly dull lecture about triangles.

The contest was won by a trainee police officer named Clare Southern, who managed to remain awake for just over a week. To stop herself falling asleep, Clare sang to herself, tensed her feet until they became painful, and prevented herself from urinating. When I met the contestants during the show, the effects of their sleep deprivation were obvious. Almost all reported feeling in a weird vegetative state and some had started to hallucinate, with one contestant already convinced that he was the prime minister of Australia.

The suffering endured by those who have attempted to remain continuously awake for several days vividly demonstrates humanity's deep-seated need for sleep. Everyone involved in these long-distance awake-a-thons has suffered a variety of negative experiences, including aggressive mood swings, extreme lethargy, paranoia, blurred vision, slurred speech, and a general sense of confusion. Indeed, Guinness World Records no longer has a category for sleep deprivation because of the potential physical and psychological problems associated with such extreme wakefulness.

Sleep deprivation and torture

In the middle of the sixteenth century the Church of Scotland made it illegal to be a witch, and during the next 200 years more than 1,000 people were executed because they allegedly practised witchcraft. One of the popular means of eliciting confessions was referred to as 'waking the witch', and involved forcing those accused of witchcraft to stay awake for several days. Tired and prone to hallucinations, these sleep-deprived women usually confessed to being a witch and often reported strange experiences, such as flying or transforming into an animal.

It would be nice to think that this primitive and barbaric method of interrogation has become consigned to the history books. Unfortunately, that is not the case.

In 2009, Barack Obama made public several top-secret government memos that had been issued under the Bush administration a couple of years earlier. The memos outlined the types of techniques that the CIA could use to interrogate suspected terrorists at secret detention centres around the world. One of the memos explicitly gave the green light to sleep deprivation, allowing interrogators to make prisoners stand against a wall, and then shackle their feet to the floor and their hands close to their chins. If the prisoner fell asleep they would start to fall down and be jolted awake by their chains. Records revealed that a number of detainees experienced this deprivation for several days and, in one instance, it was maintained for more than a week.[13]

In addition to the obvious ethical considerations, the procedure may well have caused the detainees to unwittingly produce unreliable information. In the mid-1990s, psychologist Mark Blagrove from Swansea University investigated the relationship between sleep deprivation and suggestibility.[14] In one study, Blagrove had a group of volunteers listen to a short description of a bank robbery. The researchers then randomly split the volunteers into two groups, and arranged for them to either sleep normally for a couple of nights or remain awake for forty-three hours. All of the volunteers were then interviewed about the crime.

The volunteers who had remained awake for just less than two days proved especially suggestible, frequently making up details to please the interviewer: evidence that sleep-deprived detainees are more likely to tell interrogators what they want to hear, rather than reveal the truth.

An accident waiting to happen

Of course, it could be argued that forcing yourself, or being forced, to spend several days and nights wide awake is a highly artificial situation, and that losing just an hour or so of sleep each night would not be especially detrimental. More than fifty years of scientific research into sleep has examined this issue, and the results are as surprising as they are terrifying.

On 24 March 1989, one of the world's largest oil tankers left Alaska with a full cargo bound for Long Beach, California. A short time into the journey the third mate attempted to alter the vessel's course to avoid several large ice floes. Unfortunately, the tanker failed to respond to the changes because the autopilot had been left on. Despite two warnings from lookouts, the third mate realized his error far too late, and was unable to prevent the ship hitting an underwater reef and spilling hundreds of thousands of barrels of crude oil into the sea.

The Exxon Valdez disaster destroyed the surrounding wildlife and habitat, and is widely considered to be one of the worst environmental catastrophes in modern history. Accident investigators concluded that sleep deprivation had played a major role in the incident. The third mate had only slept six hours in the previous two days, and was simply too tired to notice that the ship was not changing course, or to respond to the warnings from the lookouts. Unfortunately, this is not a unique event. Similar investigations have revealed that sleep deprivation played a key role in several other catastrophes, including Three Mile Island, the Challenger space shuttle disaster, and the Chernobyl meltdown.

You don't have to be in charge of a tanker, nuclear plant, or space programme to fall foul of the effects of sleep deprivation.

In fact, even a small amount of sleep deprivation can dramatically increase the chances of you having a serious accident in everyday life.

Gregory Belenky has dedicated much of his career to understanding the psychological dangers of being half asleep. For almost thirty years Belenky worked for the US Army Medical Corps and led the military's fight against fatigue. Belenky assessed many different ways of keeping soldiers awake on the battlefield, including a movement-sensitive wristwatch that can detect sleep patterns and special caffeine-impregnated chewing gum that can deliver the punch of a double espresso within seconds.[15] Belenky eventually decided to swap his camouflage jacket for a lab coat, and now has the altogether more sedate task of heading up the Sleep and Performance Research Center at Washington State University.

In 2003, Belenky and his colleagues staged one of the world's most comprehensive studies into sleep loss and vigilance.[16] The researchers asked volunteers to spend two weeks in a sleep laboratory. For the first few days the team ensured that all of the participants were tucked up in bed by 11 p.m. and allowed to sleep until 7 a.m. The researchers then randomly allocated the volunteers to one of four groups, and ensured that each group obtained a different amount of sleep each night. The volunteers in one of the groups were lucky enough to spend a luxurious nine hours in bed, while those in the other three groups were allowed to sleep either seven, five, or three hours per night.

For the next few days the researchers asked the volunteers to rate how tired they felt, and tested their alertness by asking them to press a button the moment they saw dots appear on a computer screen. The volunteers who obtained nine hours' sleep each night remained highly alert, while those spending just three or five hours in bed quickly became tired and inattentive. However, the results from those getting seven hours' sleep per night proved especially surprising. Although these volunteers constantly assured the researchers that they were as wide awake as those spending nine hours in bed, the data from the 'push the button when you see the dot' test revealed a very different story. After just a couple of days of getting seven hours' sleep they became significantly less vigilant, and remained sluggish for the remainder of the experiment.

Belenky's study revealed the highly pernicious nature of even a small amount of sleep deprivation. Spend just a few nights sleeping for seven hours or less and your brain goes into slow motion. To make matters worse, you will continue to feel fine and so don't make allowances for your sluggish mind. Within just a couple of days this level of sleep deprivation transforms you into an accident waiting to happen.

The situation becomes potentially deadly whenever you climb into your car, with sleep deprivation slowing reaction time and causing a strange form of blackout known as a 'micro-sleep' (see overleaf). The scale of the problem is terrifying. The US National Highway Traffic Safety Administration believes that such fatigue causes more than 100,000 road accidents and 1,500 fatalities each year. The issue is especially acute in teenage drivers, with drowsiness being the number one cause of fatal car crashes for eighteen- to twenty-five-year-olds. It seems that no teenage driver is immune from the problem, with a boy who'd won the accolade 'America's Safest Teen Driver' dying after falling asleep at the wheel and drifting into oncoming traffic.[17]

Worryingly, the high number of fatigue-related accidents is just the tip of the sleep deprivation iceberg.

The mystery of micro-sleep

When you are sleep deprived your brain can nod off for a few seconds without you being aware of it. Amazingly, your eyes can remain open, even though your brain is sound asleep. It is as if the lights are on, but no one is at home. This mysterious phenomenon is known as micro-sleep and can happen at any time during your waking day. If you happen to be reading at the time you may suddenly realize that you have no idea what the last sentence was about. If you are chatting then you will feel as if you have lost part of a conversation for a few moments. If you happen to be driving at the time, you may be moments away from death.

In 2012, reporter Ron Claiborne from ABC News teamed up with medical researcher Charles Czeisler to investigate the impact of micro-sleeps on driving.[18] Claiborne stayed awake for thirty-two hours to mimic the effects of chronic sleep deprivation. The researchers then connected Claiborne to a device that measured his brain activity, put him at the wheel of a minivan, and asked him drive around a closed test track for a couple of hours. Claiborne appeared to be awake, and his eyes remained open for almost the entire trip. However, his brain activity revealed that that he took more than twenty micro-sleeps during the drive.

Although micro-sleeps only last a few seconds, they can easily cause drivers to shoot across red lights or drift into oncoming traffic. Investigators are able to spot road accidents caused by micro-sleep because there is usually little evidence of the driver attempting to respond to an obvious hazard.

> Fatigued drivers don't hit the brakes or try to swerve. Instead, they simply drive directly into danger while sound asleep.

Sleeping on the job

Your brain makes up just 2 per cent of your weight but uses 20 per cent of all the energy your body produces. When you are sleep deprived your body struggles to extract glucose from the bloodstream, and so your brain cannot think straight.[19]

To investigate this effect, sleep scientist Jim Horne from Loughborough University once created his own mini-casino and ensured that the only game in town was extremely simple.[20] Volunteers were invited to Horne's laboratory-cum-casino, and asked to pick cards from four decks. Some of the cards earned them money, and others lost them money. The decks were stacked so that they contained different proportions of winning and losing cards. The volunteers who had got a good night's sleep unconsciously learnt to choose cards from the decks with a high proportion of winning cards, whereas those who were sleep deprived continued to select randomly from all four decks. Given Horne's findings, it isn't surprising that many casinos encourage people to spend the night gambling rather than head for their beds.

But sleep deprivation doesn't just affect your ability to make rational decisions. The front part of your brain plays a key role in determining your level of willpower. When you are sleep deprived, the low energy levels in this part of the brain can be especially damaging because it disrupts your sense of self-control and discipline. Several studies have revealed the damaging impact that this can have in the workplace.

In one experiment, Michael Christian from the University of North Carolina at Chapel Hill examined the relationship between sleep deprivation and professionalism in hospitals. Christian asked a group of nurses to indicate how many hours they had slept the previous night and the degree to which they had engaged in various deviant behaviours during their last shift, such as intentionally working slowly, having inappropriate discussions about confidential information, and using illegal drugs on the ward.[21] The nurses that had obtained less than seven hours' sleep engaged in significantly more unprofessional behaviour.

Another study, carried out by Christopher Barnes from the Pamplin College of Business at Virginia Tech, concentrated on cheating.[22] Barnes asked a group of students to keep a sleep diary for a week and then take part in a trivia quiz. The students were given the opportunity to cheat by altering their answers, and so the researchers were able to discover whether less sleep resulted in more cheating. The later the students had gone to bed, the more they cheated.

Finally, David Wagner from Singapore Management University examined whether sleep-deprived employees were especially likely to surf the Internet for fun while they were at work (known in the trade as 'cyberloafing').[23] Wagner first fitted volunteers with special wristwatches that used accelerometers to measure how many minutes they slept during the night. The following day the volunteers were asked to watch an important video lecture on their office computers. Unbeknown to them the researchers secretly monitored the number of minutes each volunteer spent visiting entertainment websites instead of watching the video. The volunteers that had had tossed and turned throughout the night were especially likely to cyberloaf the following day.

The effect of such low self-control in the workplace is far from trivial, with sleep-related fatigue costing businesses an estimated $150 billion a year in lost productivity.[24]

Worse still, over the long haul these effects can take a terrible toll on your brain. In one study, for example, Jane Ferrie, a public health researcher at University College London Medical School, studied the lives and minds of more than 5,000 middle-aged volunteers over a five-year period.[25] In the late 1990s, all of the volunteers indicated the average amount of time they slept each night. Five years later they all reported the amount of sleep they were getting a second time, and also completed tests measuring their memory, vocabulary, and logical thinking. The results revealed that sleeping for less than six hours each night, or much more than eight, was associated with lower test scores. In addition, those volunteers who showed a significant decrease in their sleep over the five-year period had especially low scores in logical thinking and vocabulary. Generally, our ability to remember information and think quickly both decline as we get older. Based on this work, researchers estimate that for middle-aged adults, sleeping less than six hours, or more than nine, ages your brain by around seven years.

Sleep deprivation is wrecking our educational institutions, costing business billions, and prematurely aging our brains. Perhaps most worrying of all, it can even kill you.

The dangers of long sleep

The vast majority of research into sleep and well-being has examined the downside of not spending enough time in bed. However, other work has demonstrated that when it comes to sleep, as with so many things in life, it's also possible to have too much of a good thing.

Several large-scale studies have shown that spending nine or more hours asleep each night is associated with a range of medical conditions, including diabetes, obesity, headaches, cancer, and heart disease.[26] In addition, some long sleepers suffer from a condition known as 'hypersomnia'. As well as spending a lot of time in bed, they tend to be extremely sleepy during the day, do not feel any more alert after they have taken a nap, are highly anxious, feel constantly tired, and experience memory problems.

Sleep follows the 'Goldilocks principle'. In the same way that one bowl of porridge was too hot and another too cold, so it's possible to spend too long and too little in bed. For most people, about eight hours asleep each night is just right.

I'll sleep when I'm dead

Starting in the mid-1980s, researchers from University College London spent twenty years examining the relationship between sleep patterns and life expectancy in more than 10,000 British civil servants.[27] The results, published in 2007, revealed that participants who obtained two hours less sleep a night than they required nearly doubled their risk of death. In a similar study, another group of researchers analysed data from more than one million Americans and found that getting less than seven hours' sleep each night was associated with an early demise.[28]

Further work has started to uncover why poor sleep kills.

Some of this research has focused on the relationship between the sleep-inducing hormone melatonin and the circulatory system. Several studies have revealed that melatonin has several positive effects on the body, including lowering blood pressure, and helping to prevent heart attacks and strokes.[29] Sleep deprivation goes hand in hand with lower levels of melatonin production, and so increases the risk of these problems. The effects are far from trivial, with research published in the academic journal *Sleep* showing that the risk of high blood pressure was more than three times greater among those who sleep for less than six hours per night, and that women who obtain less than four hours of sleep are twice as likely to die from heart disease.

But melatonin doesn't just keep the blood running through your veins, it also limits the production of other hormones that are associated with various cancers. The impact of failing to produce these hormones can be dramatic, with large-scale studies showing that female shift workers (who frequently suffer from low melatonin levels) have a 60 per cent increased risk of breast cancer[30] and a 35 per cent greater risk of colorectal cancer.[31] Other work suggests that poor sleep is also related to the onset of diabetes, with research showing that people who sleep for less than five hours per night are more than three times more likely to develop type 2 diabetes.[32]

There is also the issue of obesity. More than 28 per cent of American adults get less than six hours of sleep a night,[33] and more than 35 per cent of the population is obese.[34] Strange as it may sound, the obesity epidemic may be fuelled by both adults and children failing to get enough sleep.

In 2005 British researchers published the results of a study that had tracked the sleeping habits and weight of more than

8,000 children from birth. The results revealed that those who slept for less than ten-and-a-half hours each night when they were three years old had a 45 per cent higher risk of being obese when aged seven.[35] Similarly, in America, researchers monitored the sleep patterns and weight of more than 60,000 middle-aged American women for sixteen years.[36] Those getting five hours or less sleep each night were 15 per cent more likely to become obese. In yet another study, researchers at Stanford University examined the relationship between people's sleeping habits and their weight.[37] The data from more than 1,000 volunteers revealed that sleeping less than eight hours a night was associated with obesity.

Of course, many of these data patterns could be due to obese people struggling to sleep. However, other research shows that a lack of sleep does indeed result in weight gain. Two hormones play key roles in regulating the amount of food you consume. One of them, known as 'ghrelin', is produced in your gastro-intestinal tract and stimulates your appetite. The other, 'leptin', is created in fat cells and tells your brain that enough is enough. Researchers at the University of Chicago monitored the hormones and appetite levels of male volunteers after they had slept only four hours a night for two nights. Remarkably, the team discovered that this short-term sleep deprivation had a major impact on their hormones and resulted in a 28 per cent increase in ghrelin and an 18 per cent decrease in leptin. These changes stimulated the volunteers' appetites but ensured that they didn't feel especially satisfied after they had eaten. As a result of this double hormonal whammy, the volunteers reported a 24 per cent increase in appetite, and were especially attracted to unhealthy calorie-dense foods, such as sweets, biscuits, and crisps.[38]

Neuroscientist Colin Chapman, from Uppsala University in Sweden, recently examined the impact of sleep deprivation on shopping habits.[39] Chapman and his team recruited a group of healthy men and staged a two-day study. During day one, the volunteers were asked to stay awake for a night, then go food shopping the following morning. On the second day they were allowed to sleep as usual and then head off to the shops again. Before going shopping each volunteer was given about £30, shown a list of forty items, and asked to buy as many of the items as possible. Half of the items were junk food and half were far healthier. When they were sleep deprived, the volunteers returned with significantly more junk food. Sleep deprivation isn't just bad for your brain – it can also ruin your waistline.

Beauty sleep

Human skin looks smooth and firm because of a protein known as collagen. As you age, the amount of collagen in your body decreases, causing your skin to become wrinkled and saggy. Several nights of poor sleep causes your body to release a stress hormone called cortisol. This hormone prevents the production of collagen, and so creates unhealthy looking skin, fine lines, and dark circles under your eyes.

These effects can take hold surprisingly quickly. In one study conducted at the Karolinska Institute in Stockholm, researchers photographed volunteers after they had just slept for eight hours and again after they had been kept awake for thirty-one hours.[40] Different volunteers were then shown the photographs and had to rate how healthy and attractive the individual looked. The photographs of the sleep-deprived volunteers were judged as less healthy and less attractive. Other work delved deeper into the effects, revealing that sleep deprivation resulted in redder eyes, an increase in dark circles under the eyes, paler skin, and more wrinkles.[41]

The message is clear: When it comes to poor sleep, the evidence is written all over your face.

Getting up and feeling down

A lack of sleep doesn't just affect your physical health. It may also impact on your psychological well-being.

Psychiatrists have long noticed that people suffering from a wide range of mental problems do not sleep well. For instance, around 90 per cent of depressives report struggling to fall asleep or waking up in the middle of the night. People with bipolar disorder experience episodes in which they are overly energetic and frenzied (referred to as 'mania'), alternating with instances in which they feel down and depressed. Again, the disorder is associated with sleep problems, with those experiencing mania often getting only three hours of sleep per night and some even going several days without sleeping.

Schizophrenia can involve a range of symptoms including hallucinations, delusions, and confused thinking. There is a strong association with poor sleeping for this too, with research showing that around 70 per cent of those suffering from the disorder struggle to fall asleep, or sleep for an especially long time, and have irregular body clocks that are out of sync with daytime and night-time.[42]

Finally, there is a curious association between sleep issues and attention deficit hyperactivity disorder (ADHD) in children. For many children, sleep deprivation does not cause lethargy but instead promotes hyperactivity and unfocused thinking. There is now evidence from several studies showing that many children diagnosed with ADHD also suffer from a breathing-related sleep disorder known as 'sleep apnoea' (more about this in Lesson 4) and have low levels of deep sleep.[43]

Of course, correlation does not mean causation, and it's quite possible that a psychological disorder could cause a sleep problem rather than the other way around. To help resolve the complex relationship between sleep and mental health, several researchers have started to help sufferers get a better night's rest and then monitored the effects. This work uses a variety of techniques, including altering people's circadian rhythms by bright-light phototherapy and taking melatonin supplements, and encouraging people to sleep by exercising during the day. Although only in its initial stages, the work has already yielded promising results. The interventions helped relieve symptoms of both depression and bipolar disorder. If valid, these findings suggest that sleep problems play a key role in causing various psychological disorders.

Sleep: The ultimate detox?

Keeping your brain in tip-top shape is a time-consuming and complicated business.

Your brain cells continually produce a kind of 'toxic waste' that, if allowed to build up, disrupts your thoughts, behaviour and mood. To avoid this happening, your body regularly flushes a special cleaning liquid called 'cerebrospinal fluid' (CSF) through your brain tissue. CSF removes the unwanted waste and carries it down to your liver for detoxification. Neuroscientists from the University of Rochester Medical Center recently made a remarkable discovery about the role of sleep in this vital waste disposal process.[44]

The researchers took a group of mice (mouse brains are remarkably similar to human brains), placed a special fluorescent dye into the CSF, and used a high-powered scanner to monitor the fluid when the animal was awake and asleep.

When the mice nodded off their brain cells shrunk, and the space between the cells increased by around 60 per cent. This allowed lots of the CSF to flow quickly through the brain and wash away the toxins in a highly efficient way. This groundbreaking work suggests that when you fall asleep, your brain becomes far more efficient at removing the unwanted toxins that build up during the day. On the flip side, if you fail to get enough sleep, these toxins remain in your brain, causing you to feel foggy-headed and irritable.

Time for a test

Many people think that not getting a good night's sleep makes them grumpy. In reality, the effects of sleep deprivation are far more damaging. Even small amounts of sleep deprivation play havoc with your brain. It may make you accident-prone, or drive dangerously, or be unproductive. A lack of sleep is also affecting our nation's youngest minds, causing them to struggle in school and fail at college. Over time, it can even prematurely age your brain. Perhaps most important of all, sleep deprivation is bad for your health. As we've seen, poor-quality sleep dramatically increases the chances of you suffering from high blood pressure, cancer, diabetes, and obesity. The problem is especially pernicious because, unlike many other psychological and physical problems, it is easy to struggle through your life without realizing the full extent of the deprivation. In fact you may be sleep deprived right now, without even realizing. Which leads us to one important question . . . are you getting enough high-quality sleep?

Over the years sleep scientists have devised various question-naires for measuring both the quality and quantity of your sleep. In fact, you completed one of them just before the start of this lesson (see page 48).

Take a look at your responses. To score the questionnaire, simply add up the numbers below the options that you circled (assuming that you didn't nod off halfway through). A score of seventeen or below suggests that you are experiencing poor quality sleep. If you did obtain a low score then you are not alone. When I asked 3,000 people to complete this questionnaire a remarkable 20 per cent of respondents fell into this category. If you scored between eighteen and twenty-six then this suggests that the quality of your sleep is middling. But does that mean that you can rest on your laurels? Absolutely not.

When I looked at the data from the questionnaire I realized that only about 10 per cent of people obtain a score of twenty-seven or more. These 'super-sleepers' are quite remarkable. They enjoy a good night's sleep almost every night of their lives, are able to fall asleep whenever they want, wake up feeling refreshed, and have lots of sweet dreams. This impressive ability to make the most of the night results in them being especially happy, success-ful and healthy. Compared to those obtaining middling scores, these super-sleepers are about 25 per cent happier, 30 per cent more able to achieve their goals (such as losing weight and stop-ping smoking), and 40 per cent less stressed. If you obtained a middling score on the questionnaire then you are just a short hop and a jump away from becoming a super-sleeper.

I believe that everyone has the ability to make the most of the missing third of their life. Some can move from being a poor to good sleeper, and others can go from good to great. All you have to do is understand the science of sleep and dreaming, and dis-cover how best to apply it to your life. In the next Night School lesson we will be taking the first step on that all-important journey.

Meanwhile, sleep well.

The national sleep deprivation map

I recently carried out a national survey examining sleep depriv-ation across Britain. More than 3,000 people were asked to indicate how much sleep they obtained on a typical weekday. Overall, the results show that sleep deprivation is a consid-erable problem in bedrooms across the nation, with around 40 per cent of respondents indicating that they spent seven or less hours asleep each night.

The problem is especially severe among people living in Yorkshire and the Humber, the North West, the South West, and London. In contrast, those in the North East and the East Midlands get the nation's best sleep.

Map reflecting the percentage of people in each region sleeping seven or less hours each night.

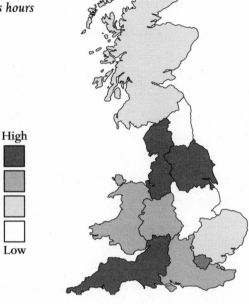

High

Low

%	Region
46	Yorkshire and the Humber
45	North West
44	London
44	South West
40	West Midlands
40	South East
40	Wales
39	Scotland
39	East of England
37	North East
26	East Midlands

ASSIGNMENT

What does your sleeping position reveal about you?

'The king sleeps on his back, the sage on his side
and the rich man on his belly'
Old proverb

Please take a few moments to answer the following question. Which of these options best describes the most comfortable position for you when you fall asleep?

1) *Full-fetal*
You lie on your side with your body almost completely curled up. Your legs are flexed at the knees, and your knees are drawn up towards your chin. Often your entire body will be rolled into a ball and may curve around an object such as a pillow.

2) *Semi-fetal*
You lie on your side with your knees drawn part way up.

3) *Royal*
You lie on your back.

4) *Prone*
You lie face down on your bed,
often with your arms over your head
and legs stretched out with the
feet somewhat apart.

In the late 1970s, psychiatrist Samuel Dunkell presented the world with a new and exciting theory of personality. In his groundbreaking book, *Sleep Positions: The Night Language of the Body*, Dunkell argued that it was possible to gain a genuine insight into a person's psyche on the basis of their preferred sleeping position.[45] Many of Dunkell's wilder assertions are fun to mention at parties. For example, people who spend the night with their legs crossed at the ankle apparently struggle with relationships, those adopting the 'flamingo' pose are allegedly passive-aggressive, and those who place their hands behind their head and point their

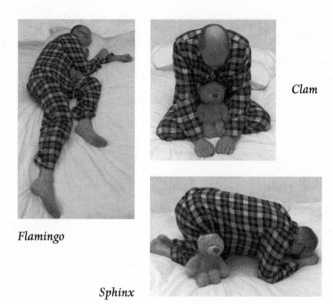

Clam

Flamingo

Sphinx

elbows up towards the ceiling seem to enjoy a good argument. Dunkell also argued that those who exhibit the rarely seen 'Clam' position are self-absorbed, while people sleeping in the 'Sphinx' position are especially strong-willed (and, I suspect, also suffer from intense lower back pain).

Over the years several sceptical academics have examined whether Dunkell's general thesis holds water. Some of the initial work investigated whether people even have a preferred sleeping position, and involved researchers asking people to repeatedly report how they laid in bed over a six-month period. The results revealed that the vast majority of them did indeed have a stable preference, and tended to adopt one of the four positions shown in the questionnaire. Excited by these findings, other scientists systematically investigated whether these positions were associated with certain types of personality.[46] To the surprise of many sceptics, the results revealed the following associations:

Full-fetal: People adopting this position tend to be anxious, emotional, indecisive, and overly sensitive to criticism. Dunkell interpreted the 'closed' nature of this position as indicative of a person who does not want to open themselves up to life.

Semi-fetal: This position is associated with people who are well adjusted, conciliatory in nature, amenable to compromises, and unlikely to take extreme stances.

Royal: This sleeping position is associated with being self-confident, open, expansive, and sensation-seeking.

Prone: Those sleeping face down tend to show a tendency for rigidity and perfectionism. Dunkell thought that these sleepers disliked the unexpected, demand strong evidence for any assertion, and always arrive on time for meetings.

The research also showed that those who have no preferred sleeping position have a strong need for being active, enjoy challenging work, and find it difficult to relax.

However, please don't be too upset or worried if your sleeping position suggests that you have a less-than-perfect personality. The associations between people's sleeping positions and their personalities are fairly weak and many scientists would take them with a pinch of salt. I suspect that this is especially true of those researchers who tend to sleep in a prone position.

Lesson 3

THE SECRET OF
SUPER-SLEEP

Where we find out the truth about 'short-sleep', discover
how to get the best night's sleep of your life,
and learn how to sleep like a baby.

In the previous lesson we found out how twenty-four-hour media, increased workloads and constant access to the web have resulted in a world that rarely sleeps. This chronic lack of rest is playing havoc with our minds and bodies. Fatigue is responsible for thousands of road accidents each year, costs businesses billions in lost productivity, and increases the risk of obesity, diabetes, and death. Because of this, there is now a pressing need for everyone to get their full quota of sleep each night.

Even if you don't have a sleep problem, improving the quantity and quality of your sleep can radically change your life. As we saw in the last lesson, research has revealed the existence of 'super-sleepers' – people who can fall asleep whenever they want, enjoy the most pleasant of dreams, wake up feeling totally refreshed, and are especially happy, productive, successful, and healthy. I believe that almost everyone can make the most of the night by learning to become a super-sleeper.

Many people react to this information in one of two ways. Some wonder if they can cheat sleep by staying fully functional but spending a small amount of time in bed. Others become eager to improve both the quantity and quality of their sleep.

In this lesson we are going to tackle both of these issues. In the first part, you will discover whether there really are ways of thriving on a small amount of sleep, and in the second part you will find out how to get the best night's sleep you have ever had. We begin by playing chess with death.

Life on the ocean waves

At the start of Ingmar Bergman's classic film *The Seventh Seal*, a knight is visited by a pale-faced figure wearing a black cowl. The figure mournfully explains that he is Death, and that he has come to take the knight's life. Understandably, the knight is less than delighted at the news, and challenges Death to a chess match in an attempt to avoid his demise.

Over the years, some people have taken a similarly sly approach to the night, and tried to figure out ways of avoiding spending a third of each day sleeping. Perhaps the best known of these 'we can all survive on very little sleep' theories revolves around research conducted by top sleep scientist, and highly experienced yachtsman, Claudio Stampi.[1]

Attempting to single-handedly sail across the world's oceans requires very high levels of vigilance, and even the smallest amount of ill-timed sleep can have disastrous consequences. In 1982, for example, solo sailor Desmond Hampton chartered Sir Francis Chichester's last yacht, *Gipsy Moth V*, to compete in an around-the-world race. During the race Hampton became exhausted, went below deck to grab an hour's rest, but promptly overslept. The *Gipsy Moth V* eventually wedged itself into rocks just off Australia's south-eastern coast and was wrecked. Similarly, in 1994, French soloist Jean-Luc Van Den Heede put his head down for a five-minute nap during another global race and woke up thirty minutes later to discover that his craft had run aground.

Stampi wondered whether his knowledge of sleep science could help competitors remain awake for long periods of time, but nevertheless feel well rested and alert. As we discovered in Lesson 1, research into circadian rhythms suggests that humans have developed to take most of their sleep at night with potential for a short nap during the afternoon (a pattern that sleep scientists refer to as 'biphasic'). This is unusual in the animal kingdom, as many other creatures take lots of short naps throughout the day. Most biologists believe that this more scattergun approach to sleep, known as 'polyphasic' sleeping, developed because animals frequently face an energy crisis. In order to survive, they have to run around eating food and avoiding enemies. Unfortunately, this high-octane lifestyle consumes lots of energy, forcing them to take frequent naps throughout the day to recharge their batteries.

Stampi realized that polyphasic sleep would be perfect for his long-distance sailing chums, and began working with yachtsmen to discover if they could remain healthy and alert on this strange schedule.[2] In perhaps his most famous project, he helped British sailing star Ellen MacArthur attempt to sail single-handed around the globe. Before the race, Stampi monitored MacArthur's sleeping habits and worked out an optimal sleep schedule. During her epic journey MacArthur then wore a small watch-like device on her arm. This device constantly monitored her movements and skin temperature, and then beamed this information back to Stampi's Boston-based Chronobiology Research Institute. This data revealed that MacArthur took almost 900 naps during her voyage, with each sleep lasting an average of only thirty-six minutes. This remarkable schedule paid off, with MacArthur sailing around the world in seventy-one days and fourteen hours, and so breaking the former record by an amazing thirty-two hours.

Stampi's research proved that it is indeed possible for humans to survive on polyphasic sleep for a few months. But could this strange form of multi-napping become a way of life?

In the 1950s, Italian artist and actor Giancarlo Sbragia heard that Leonardo da Vinci was amazingly productive – in part because he slept for fifteen minutes every four hours and in doing so gained an additional six hours of working time each day. Sbragia reasoned that if this polyphasic sleep schedule was good enough for one of the world's greatest artists, it was good enough for him, and decided to give it a whirl. Sbragia initially struggled to keep his eyes open, but after a couple of weeks he was able to get through the day by taking fifteen-minute naps six times a day. Initially, the punishing schedule gave him more time to indulge his artistic passions, and all seemed well. However, as the months rolled on, Sbragia ran out of activities to fill his time and this, coupled with the lack of around-the-clock company, led to feelings of intense loneliness. Sadly, he slowly came to the painful realization that his lack of Leonardo-like fame was not due to there only being so many hours in the day, but rather a more general deficiency in the genius department. He eventually abandoned the experiment after just six months.

Sbragia is not the only one to have tried, and failed, to live life on polyphasic sleep. Search the Internet and you will see several reports of people experimenting with all sorts of scientific-sounding polyphasic-sleep schedules, including the 'Everyman' (four-and-a-half hours of sleep at night plus two twenty-minute naps during the day), the 'Dymaxion' (four thirty-minute naps every six hours), and the 'Uberman' (six twenty-minute naps every four hours). These 'experiments' nearly always follow the same pattern. For a few weeks life is very tough, but eventually the person gets used to the schedule and begins to enjoy the novelty of the experience. However, as the months go by they slowly start to exhibit various forms of sleep deprivation, find the schedule tricky to maintain in the face of work commitments and social life, and are glad to return to their old sleeping habits.

Stampi's work is fascinating, and shows that humans are able to maintain polyphasic sleep schedules for a relatively short period of time. However, time and again research has shown that this schedule isn't a replacement for a good night's sleep and can't be maintained over the long haul. Is this failure the final nail in the coffin of short sleep? Actually, no. Enter psychologist and radical sleep scientist Ray Meddis.

Short sleep

In the 1970s psychologist Ray Meddis came up with a radically new theory of sleep.[3] According to Meddis, sleep evolved to help animals remain inert and inconspicuous when they are most at risk from predators. By lying totally still in the dark, a sleeping animal is far less likely to be attacked than their 'It's dark outside and I am going out for a wander. I mean, what's the worst that could happen?' counterparts. For Meddis, sleep has very little to do with keeping the brain and body refreshed, and is instead all about staying alive. Perhaps most contentious of all, he also argued that for most people sleep now serves no real purpose because the chances of them being attacked by a tiger in the middle of the night are, at best, slim. Meddis's theory predicted that it should be possible for humans to live life to the full on very little sleep. To investigate, he set off in search of people who felt fine despite spending very little time in bed.

Throughout history, many famous figures have claimed that they are fully functional despite only sleeping for a few hours each night. However, dig a little deeper and you will quickly discover that some of these claimants are being a tad economical with the truth. For instance, the world-famous inventor of the electric light bulb, Thomas Edison, repeatedly told journalists that he only required a few hours of sleep each day, and once proudly declared: 'I never found need of more than four or five hours' sleep in the twenty-four. I never dream. It's real sleep.' However, Edison's biographers have discovered that the great inventor enjoyed more than his fair share of sleep. Edison had beds installed in his office and library, as well as several other rooms across his property. The great inventor would frequently lie down and nap, resulting in him obtaining far more than four or five hours of sleep each day.

Wary of the possible apocryphal nature of historical reports, Meddis spent several years trying to track down people who were happy and healthy, but yet required very little sleep each night. Requests in newspapers and on television produced lots of potential leads, but many of the respondents were either suffering from insomnia (in which case they weren't especially happy and healthy) or kidding themselves. In one instance, Meddis was contacted by a middle-aged man who claimed that he hadn't slept a wink since he was fifteen years old. When the man was brought into the laboratory he slept like a baby for almost six hours.

After years of searching, one of Meddis's friends introduced him to an energetic seventy-year-old retired nurse who claimed to sleep for less than an hour a night. Intrigued, Meddis and his colleagues decided to put this remarkable claim to the test.[4] To preserve the woman's anonymity, the team only ever referred to their participant as the mysterious 'Miss M.' in their scientific papers.

The researchers began by trying to monitor Miss M. for a few days to find out whether she really did get by on very little sleep. On the first night they invited her to their sleep laboratory in the hope of connecting her to an EEG machine. Unfortunately, the team had underestimated the curiosity and drive of their elderly participant. Fascinated by the high-tech equipment and delighted with her newfound companions, Miss M. refused to go to sleep and merrily chatted the night away. When the morning rolled around the researchers were shattered and so had to take turns observing her throughout the following day. Once again, Miss M. was more than happy to chat and spent another twenty-four hours wide awake. On the third night the researchers finally persuaded Miss M. to go to bed. She slept for just one-and-a-half hours before getting up and continuing to chat.

Fascinated with these initial results, Meddis extended his around-the-clock surveillance of Miss M. for another week. During the day, groups of 'mature girl students' carefully observed her as she went about her daily business, and each night the energetic seventy-year-old spent time at the sleep laboratory. The results revealed that Miss M. was indeed a genuine short-sleeper, averaging just over an hour of sleep each day. Delighted to find the real deal, the researchers examined Miss M.'s EEG records in the hope of discovering why she was able to survive without sleep. Unfortunately, the EEG data proved of little use, and just confirmed that her sleep contained the same percentage of NREM and REM episodes as those spending eight hours a night in bed.

Over the years researchers have examined the brains, bodies, and lives of other people who spent very little time in their bed. Some of the most recent work has shown that they tend to be a highly energetic group, with a surprisingly large percentage enjoying careers as high achieving CEOs and entrepreneurs.[5] Some of the most interesting work into the topic has been carried out by geneticists from the University of California. In 2009, the researchers noticed something very odd when they were examining the DNA in blood samples from volunteers who had participated in a series of sleep studies.[6] Rather unusually, two of the samples had abnormal copies of a gene called 'DEC2', which is known to affect people's sleep schedules. Curious, the researchers tracked down the participants and were amazed to discover that the samples came from a mother and daughter who were both short-sleepers. Not only that, but when other scientists bred mice with the same genetic mutation, the animals slept less and showed very few signs of sleep deprivation. This work strongly suggests that the ability to spend the vast majority of each day wide awake is due to an unusual genetic anomaly, explaining why it usually starts in childhood and runs in families. As a result, in the same way that people with black hair cannot 'train' themselves to become blonde, so the vast majority of the population will never be able to learn how to fully function on just a few hours of sleep each night.

The conclusion from research into polyphasic sleep and short-sleepers is clear. When it comes to sleep, there are no quick fixes. As a result, it is vital that everyone knows how to get a good night's rest. Luckily, several sleep scientists have dedicated their careers to finding out the best ways of getting to sleep, and staying asleep. And that's where we are heading next.

How much sleep do you need?

According to textbooks, most adults need to sleep for between seven and eight hours a day to remain fully functional. As a rule of thumb, it isn't a bad estimate. However, everyone also varies in exactly how much sleep they actually require. The good news is that scientists have created several procedures to determine how much sleep you need each night. Here are two of the simplest:

The wake-up call
This exercise takes about two weeks. First, choose a convenient time to wake up each morning. You are free to choose any time at all, but ensure that you are unlikely to get woken up earlier, and that you are happy to wake up at your chosen time during the week and at weekends. Second, wake up and force yourself out of bed at your chosen time each day (even if it means putting your alarm clock on the other side of your room). Finally, only head to bed each night when you actually feel tired. Don't force yourself to go to bed if you are not tired, and don't stay up if you feel sleepy. After a couple of weeks your body will learn that it has to get up at your chosen time in the morning, and will adjust your level of night-time tiredness to ensure that you get the sleep you require. The number of hours that you spend in bed is the amount of sleep you need.

Running free
This requires more effort than the first procedure, and involves a process known as 'free-running'. You will need to

find a week that you can dedicate to sleeping, and so it might be best to carry out the exercise when you have some time off work. Turn off your alarm clock, choose a typical bedtime and allow yourself the luxury of sleeping as long as you want. After a day or two you will have paid off your sleep debt, and you will start to wake up at roughly the same time each day. The average number of hours between when you fall asleep and when you are waking up is the amount of sleep that you require each day.

People are often stunned when they carry out these simple procedures, and suddenly realize just how much sleep they are missing every night of their lives.

Super-sleep

That most talented teller of tales, Charles Dickens, was obsessed with trying to get a good night's sleep throughout his entire life. The great writer tried an endless array of techniques, including ensuring that he was lying in the exact centre of any mattress and that the bed was pointing towards the north.[7] Not surprisingly, these strange procedures proved of little use, resulting in Dickens staying awake for hours on end, and often going for long night-time walks around the streets of London.

Dickens was not the only Victorian to experiment with rather unusual ways of trying to get a good night's sleep. In the late nineteenth century, small magnets, referred to as 'lodestones', were frequently sewn into pillows in an attempt to calm sleeping minds. More radical 'cures' involved placing a large number of pillows under people's feet, or having them spend the night on a top-heavy 'sleep boat' that would constantly rock them up and down. This latter approach apparently resulted in a ghastly form of seasickness that, on the upside, made people's insomnia seem relatively trivial. Other techniques were just downright bizarre, with, for instance, an 1894 edition of the *British Medical Journal* describing some extremely odd recommendations: 'Soap your head with the ordinary yellow soap; rub it into the roots of the hair until your head is just lather all over, tie it up in a napkin, go to bed, and wash it out in the morning. Do this for a fortnight.'

It would be nice to think that these types of strange and pseudo-scientific approaches to sleep have been consigned to the history books. Unfortunately, that is not the case. Search the Internet and you will find endless 'scientifically proven' ways of improving your sleep. From drinking milk to warm baths, smelling lavender to counting sheep, and eating cheese to doing press-ups, every sleep guru has their own top tips. But what really works?

In Lesson 2, I described how my research had uncovered a group of people that I referred to as super-sleepers. These people enjoy a good night's sleep almost every night of their lives and nearly always wake up feeling completely refreshed. This impressive ability to make the most of the night results in them being especially happy, successful, productive, and healthy. To find out the secret of a great night's sleep I surveyed the super-sleepers involved in my research, and then combined my findings with decades of work from sleep scientists from around the world. The results revealed that a great night's sleep comes down to following simple suggestions and techniques in five key areas. These techniques can be used by those who struggle to sleep, and those who do not have a problem per se but simply wish to enjoy higher-quality sleep.

Let's work our way through the suggestions and discover how to craft the perfect night in.*

1) *Create your own bat cave*

Bats sleep for around sixteen hours a day. In order to achieve this remarkable amount of shut-eye they live in secure caves away from potential predators. It's important to produce your own bat cave by creating a nice, quiet, dark place that helps you sleep at night. This may sound somewhat obvious but, as the French philosopher Voltaire once famously pointed out, the main problem with common sense is that it is not so common. Here is a guide to creating the perfect space for sleeping:

Embrace the darkness

As we discovered in Lesson 1, your brain produces less of the sleep-inducing hormone melatonin when your eyes are exposed to light. It's easy to underestimate the powerful effect that light can have on your brain during the evening, with research showing that being exposed to just one hour of moderately bright light at night reduces the amount of melatonin in your brain to daytime levels.[8] As a result, it's important to ensure that you use ambient lighting in your living room and bedroom, and do not suddenly turn on any especially bright lights in your bathroom before you head to bed. Finally, ensure that your bedroom is as dark as possible. Some people use heavy curtains, while others

---

* This advice is designed to illustrate techniques used by health professionals. If you believe that you, or your child, have a psychological problem, please consult a professional.

invest in eye masks. Either way, ensure that daylight is an unwel-come guest.

Although any type of light stops you feeling sleepy, research has shown that light towards the blue end of the spectrum is especially effective at keeping you awake. Unfortunately, com-puter screens, tablets, smartphones, flat-screen television, and LED lighting all emit large amounts of blue light. There are various measures that you can take to limit your exposure to blue light in the few hours before you go to bed:

- If you must use your smartphone, tablet, or computer late in the evening, try turning down the brightness, ensuring that the device is at least twelve inches from your eyes, and consider using an app that dims the lighting on your screen at night.
- Consider wearing a pair of amber-tinted glasses that are designed to block blue light. Although you might look a tad strange, research shows that this type of eyewear is highly effective at improving sleep quality and mood.[9]
- If you want to use a night light, choose one with a dim red bulb – red light tends not to suppress the production of melatonin and so will help make you feel sleepy.

The sound of silence

When you sleep, your unconscious brain is still listening out for any sounds that signal danger, such as late-night revellers, sirens, or even a door creaking open. Men and women are sensitive to different kinds of sounds when they sleep, with some research showing that woman are listening out for crying babies, dripping taps, and rowdiness, while men are more attuned to car alarms, howling wind, and buzzing flies.[10] In other noise-related news, Dutch researchers Henk Miedema and Henk Vos turned their attention to trains, planes, and automobiles. The two Henks analysed data from more than 20,000 people to discover the types of transport-related noise that are most likely to disrupt sleep.[11] After controlling for the volume of the noise, the team discovered that aircraft are more disruptive than cars, and that cars are worse than trains. However, if you do live on a flight path, by a busy road or railway, help is at hand. Other research has shown that playing the sounds of waves crashing on a beach, or 'white noise' (the noise that you hear when you have a radio tuned between stations), helps cover up these disturbances and aids sleep.[12]

The heat is off
It is important to ensure that your room is not too hot or too cold. If your sleeping environment is overly warm then you will quickly start to feel as if you are overheating, while a cold room can keep you awake all night. Most sleep scientists recommend that your bedroom is just over 18°C (65°F), and about 65 per cent humidity. Covered with a normal amount of bedclothes, your body remains 'thermally neutral' at this temperature, and so you don't have to create heat by shivering or cool down by sweating. Temperature control might be especially important for insomniacs because they tend to have a warmer core body temperature than others. Finally, beware the curse of cold feet. Blood flow distributes heat throughout your body, and if you have bad circulation, your extremities will get cold and cause sleeplessness. If this is the case, wear a pair of warm socks to bed. As the English writer William Hone noted in 1841, 'Never go to bed with cold feet, or a cold heart.'

Safe and sound
Think about how secure you feel when you sleep. Deep-seated fears about being attacked, injured, or even killed during the

night can play havoc with your unconscious mind. Consider ways of alleviating these worries by, for instance, fitting an especially secure door or window locks, or installing a new smoke detector or burglar alarm.

Think bedroom, think sleep

It is important to associate your bedroom with sleep. As a result, many sleep scientists recommend that you only sleep and have sex in your bedroom (although, presumably, not at the same time), and try to avoid working, surfing the web, or watching television in there. Consider banishing anything – such as televisions, desks, or computers – that is encouraging you to associate the room with anything other than the two Ss.

2) *What to do during the day*

Sleep is not only about what happens at night. Super-sleepers said often that many of their daytime activities were designed to help them nod off at night. Think about incorporating the following ideas into your daytime routine.

Don't over-nap

If you are struggling to sleep at night, it could be that you simply aren't very tired. Might you be spending too long napping during the day? Your circadian rhythm encourages you to take one, twenty-minute, nap towards the middle of the afternoon. If you are napping for longer than that, think about cutting down.

Let's get physical

Researchers have carried out hundreds of studies examining whether people that exercise during the day sleep especially well at night. In 2010, Matthew Buman from the Stanford University School of Medicine reviewed these experiments and concluded that certain types of exercise do indeed promote sleep.[13] However, to help ensure that you maximize your chances of nodding off at night, you need to carry out at least two-and-a-half hours of moderate aerobic activity, or at least an hour-and-a-quarter of more vigorous exercise, each week. The studies also showed that working out around six hours before your bedtime was especially good, in part because exercise can make you all hot and sweaty, and you need time to cool down before heading to bed. Finally, don't worry if you don't enjoy spending time in the gym or pounding the pavement. Some of the most recent work suggests that both yoga and t'ai chi will help you get a good night's sleep.

Busy the mind

Spend a day at the seaside and you will probably feel more sleepy than usual. Many people attribute this phenomenon to the mysterious effects of taking the 'sea air'. However, sleep scientist Professor Jim Horne from Loughborough University believes that there is another explanation. Horne thinks that people become sleepy when they are mentally, rather than physically, tired.[14] According to Horne's theory, when you take a trip to the seaside your brain becomes exhausted taking in all of the new sights and sounds, and it is this that makes you feel tired. To test his theory, Horne had one group of volunteers walk around a boring hall all day while another group went sightseeing. Even though each group expended the same amount of energy, those that were out and about ended up feeling far more tired. So, if you want to sleep well, get out of the house and spend the day window shopping, sightseeing, or visiting a new museum or gallery.

Know when to head to bed

It sounds silly, but many people fail to get to sleep because they aren't tired. One of the quickest ways of overcoming a sleep problem is to simply aim to get up fifteen minutes earlier than normal. If that doesn't work, try implementing a tough but effective regime called 'restrictive sleeping'.[15] For the first few days do not nap during the day, and then go to bed six hours before you want to wake up. Note down how much you slept during the night, and after about five days figure out your 'sleep efficiency' by dividing your average time asleep by your average time in bed. So if, for example, you had slept for five hours, but been in bed for six hours, then your 'sleep efficiency' would be 0.83. If your sleep efficiency is greater than 0.9 then go to bed fifteen minutes earlier for the next five days. If, on the other hand, your sleep efficiency is less than 0.9, then go to bed fifteen minutes later. Repeat this process until you are getting between seven and nine hours' sleep each night, but with a 'sleep efficiency' of 0.9.

Strange bedfellows

Endless nights enduring your partner's snoring, sheet hogging, differing sleep schedules, and tossing and turning can all prevent a good night's sleep. No surprise then that around a quarter of American couples now sleep in separate beds and that many new homes are being built with two master bedrooms. If your partner is keeping you up all night, think about sleeping apart (perhaps scheduling time for intimacy at the start of the night and a snuggle in the morning), buying a king-sized bed, or filing for divorce.

---

3) Just before you head for bed

The thirty minutes or so directly before you go to bed is very important to the quality of your sleep. Use this time wisely by trying some of the following techniques.

Spend time in the tub

Several sleep researchers have arranged for people to take long baths in the name of science. In one study, David Bunnell and his colleagues from the University of California had people take baths at different times during the day, and then examined how well they slept at night. The procedure was simple. Each of the volunteers had a rectal thermometer inserted into them and were then asked to carefully sit or stand on a chair (my guess is that the majority chose to stand). Next, the volunteer was lowered on the chair into a giant bath that had been heated to a very pleasant 41°C (106°F) and spent half an hour immersed in the warm water. The research team ensured that the baths were taken at different times throughout the day, and then monitored how well the volunteers slept at night. Baths taken in the morning and afternoon had almost no effect on sleep. However, those taken in the evening, or directly before bedtime, significantly improved the quality of sleep. Scientists aren't entirely sure why a bath at bedtime is so effective, but many believe that it is all about body temperature. Research shows that there is a slight decrease in your body temperature just before you fall asleep. Lying in a warm bath artificially raises your body temperature, but when you climb out of the bath this temperature abruptly drops and sends a signal to your body that you are ready for sleep.

So, if you want to get a good night's sleep, take a nice long bath. And remember that adding bath foam helps because the bubbles insulate the water and so keep the heat in for longer.

Make a list

People often struggle to get to sleep because they are worrying about a problem in their lives, or thinking about what they need to do in the morning. Can the humble pencil and paper help solve the problem? Colleen Carney from Duke University Medical Center recruited a group of volunteers with sleep problems and randomly split them into two groups.[16] Before they went to bed, each group was asked to write down a list containing at least three things that they were worried about. The volunteers in one of the groups were also encouraged to note down something they could do to help solve the problem. Everyone was then asked to fold their list in half and put it on their bedside table. Those that had thought about how they might start to solve their problems were more relaxed at bedtime. So, if you want to have a good night's sleep, keep a pad by your bedside and before you nod off make a list of what is on your mind, and how you might start to solve these problems. Alternatively, if you tend to stay awake thinking about what you have to do the following day, just use the pad to make a 'to do' list.

The science of snacking

Most people know to avoid large meals and caffeine before going to bed, but aren't aware of which types of food and drink promote sleep. The science is straightforward. First, avoid the lure of the nightcap. Although research shows that even a small amount of alcohol does indeed put you to sleep quicker, it also gives you a more disturbed night, increases the chances of snoring, and disrupts dreaming.[17] Second, you might have heard that a late-night turkey sandwich will help you sleep because it contains a large amount of a sleep-inducing amino acid called 'tryptophan'. In fact, this is a myth, with the American Chemical Society noting that even the large amount of turkey consumed by Americans at Thanksgiving would be unlikely to cause drowsiness.[18] On the upside, research shows that you can easily increase your chances of getting a good night's sleep by eating a small portion (under 200 calories) of food that is rich in carbohydrates. If you do feel like a late-night snack, go for a small amount of high-carb chow, such as a few biscuits, one slice of toast, a small muffin, a banana, or a small bowl of wholegrain cereal.

The lavender smell mob

Several studies show that a subtle whiff of lavender helps people nod off.[19] In 2008, psychologist Chris Alford, from the University of the West of England in Bristol, sprinkled either lavender or odourless almond oil on the bedclothes of female insomniacs, and discovered that the lavender helped improve the quality of their sleep. In similar work, other researchers discovered that lavender-scented bath oils, pillows, and blankets helped improve the sleep of both infants and their mothers. Try a lavender diffuser or oil to ensure that your room smells of sleep.[*]

4) How to fall asleep

Many people lie in bed struggling to fall asleep. Both super-sleepers and sleep scientists have developed lots of techniques to help people nod off the moment they hit the hay. Strange as it may sound, it's all about having happy thoughts, faking a yawn, and trying your very best to stay awake.

------

[*] Guidelines issued by the National Institute of Health do not recommend that lavender-based products are used by women who are pregnant or breast-feeding due to a lack of knowledge about its effects.

Counting sheep

According to the old adage, if you want to get to sleep, imagine an endless series of sheep leaping over a fence and keep a running count of the total number of animals involved. Unfortunately, this particular technique has never been subjected to scientific scrutiny. However, work by Stephen Haynes from the Southern Illinois University suggests that it may indeed help some people nod off.[20] Haynes asked both insomniacs and good sleepers to carry out moderately difficult mental arithmetic problems (such as counting backwards from one hundred in threes) as they tried to fall asleep. Those without any sleep-related problems took longer than usual to nod off, while the insomniacs did indeed get to sleep quicker. If you are not good with numbers, try thinking of a category (such as 'countries' or 'fruit and vegetables') and then coming up with an example of that category for each letter of the alphabet (for example, 'A' is for 'Albania', 'B' is for 'Bulgaria', or 'A' is for 'Apple', 'B' is for 'Banana').

Think happy thoughts

Other work in the same vein suggests that there may be far more pleasant, and non-sheep related, ways of getting to sleep. In one experiment, Allison Harvey, now at the University of California, randomly allocated insomniacs to one of three groups and gave each group different bedtime instructions.[21] One group was asked to imagine a situation that they found pleasant and relaxing, another group was told to try to forget about their worries and concerns, and the final group received no guidance at all. The results were remarkable. The insomniacs who hadn't been given any special instructions took over an hour to get to sleep, while those trying to forget about their concerns took just over forty minutes. However, the volunteers who had been asked to think pleasant thoughts nodded off in just over twenty minutes. To follow this technique, try creating a wonderful fantasy world in your head. Avoid any imagery that is too exciting or sexually arousing. Instead, perhaps plan your perfect holiday, imagine how you would spend a lottery win, think about a great evening out, or set off on an amazing adventure in a fantasy spaceship.

The magic yawn

One of my previous books, *Rip It Up*, describes how the way in which you physically behave influences how you feel. For instance, smiling makes you feel happy and forcing your face into a frown makes you feel sad. This is also true of sleep. When you behave as if you are sleepy you become tired. To take advantage of this strange effect, let your eyes droop, your mouth hang open, and your arms and legs feel heavy. Sink into your bed as if you have had a long and tiring day in the office. Even fake a yawn or two. In short, fool your body into thinking that it is time for bed.

The paradox

Medical researcher Niall Broomfield from the University of Glasgow wondered whether some reverse psychology could be used to help people sleep.[22] Broomfield assembled two groups of volunteers and monitored their sleep for two weeks. One group was asked to spend each night trying to stay awake for as long as possible, while the other group didn't receive any special instructions. Those trying to stay awake felt less anxious at bedtime and reported falling asleep quicker. So, if you want to fall asleep, try to stay awake! However, remember that you have to rely on the power of your mind. You are allowed to keep your eyes open, but no reading, watching television, or moving about allowed.

The power of association

The famous Russian psychologist Ivan Pavlov spent much of his life exploring the concept of association. In his most famous experiment, Pavlov rang a bell each time he presented a dog with food, and eventually found that the sound of the bell alone was enough to make the dog salivate. The same idea can be used to help you fall asleep. Choose a soporific piece of music that you like, and ensure that it's quietly playing as you fall asleep. Over time your brain will come to associate the piece of music with sleep, so simply listening to the music will help you to nod off.

5) What to do if you wake up during the night

Some people suffer from 'sleep maintenance insomnia', waking up during the night and then struggling to get back to sleep. If this happens to you, use the following techniques to get back to sleep quickly.

Get up

If you have suddenly woken up because you have remembered something that you need to do the next day, simply make a note of your thought and try to go back to sleep. However, if you do wake for more than about twenty minutes during the night, most sleep scientists recommend getting out of bed and doing some form of non-stimulating activity. Although many people read a book or magazine, Professor Jim Horne recommends an activity that you find pleasant and relaxing, and uses your hands as well as your head. In his book *Sleepfaring*, Horne advises people to work on a jigsaw or art project. Whatever it is that you decide to do, avoid bright lights and computer screens. And if the problem re-occurs later in the night, climb back out of bed and distract yourself again.

Don't worry, be happy

Perhaps not surprisingly, spending the night feeling as if you are unable to sleep makes many people feel anxious. Over the course of a night this anxiety disrupts their sleep even more, creating a vicious downwards cycle. If you do find yourself lying in bed becoming anxious because you think you aren't getting enough sleep, here are some tips.

- Remember that you are probably getting more sleep than you think. Research shows that we all tend to underestimate how much of the night we spend sleeping. In one study, for instance, Allison Harvey measured how much time insomniacs spent sleeping during the night, and then compared it to how much they thought they had slept.[23] The insomniacs were convinced that they had only slept for an average of about three hours per night, whereas in reality they had been asleep for an average of nearer seven hours. Psychologist Jeremy Mercer from Flinders University in South Australia has attempted to discover the cause of this strange phenomenon. In one study Mercer invited insomniacs to his sleep laboratory, woke them from REM, and asked them whether they had just been sleeping. Remarkably, many of the volunteers believed that they had been wide awake despite having been sound asleep, thus raising the intriguing possibility that they were essentially dreaming about being awake.[24] Whatever the explanation, other research suggests that simply showing insomniacs data proving that they had indeed had a good night's rest, causes them to feel less anxious about getting to sleep.[25]

- Just relaxing in bed is good for you, even if you are not asleep. Rather than focusing on nodding off, try a simple relaxation exercise to make the most of the downtime. One of the most effective involves starting off by tensing your toes for about ten seconds, and then letting them relax. Use the same procedure as you work your way up the body, tensing and relaxing your legs, arms, hands, chest, shoulders, and head.

• Finally, know that your night-time wakening may have its roots in our pre-industrial sleep patterns (see below) and so be perfectly normal.

When waking up in the middle of the night is good for you

During the late 1990s, Virginia Tech historian Roger Ekirch examined how people's nocturnal behaviour had changed over time.[26] While looking through historical diaries, prayer manuals, and medical books, Ekirch uncovered a stream of strange references to the notion of 'first' and 'second' sleep. Digging deeper, Ekirch discovered that many people didn't take their sleep in one solid block at night, but instead slept for about four hours (first sleep), woke up for roughly an hour, and then went back to bed for another four hours (second sleep). The work also revealed that the time between the two periods was used for various activities, including quietly thinking, reading, smoking, praying, chatting, having sex, and sometimes even visiting neighbours.

Several scientists suggested that this seemingly strange sleep schedule, now known as 'segmented sleeping', was a natural response to the long periods of darkness associated with pre-industrial winters in the northern hemisphere. To find out if this was the case, psychiatrist Thomas Wehr from the National Institute of Mental Health turned back the hands of time and had volunteers experience artificially constructed days that contained just ten hours of light and fourteen hours of darkness.[27] After a short time the volunteers naturally drifted into a segmented sleep schedule.

Other researchers have argued segmented sleep may be good for the mind. The period between people's first and second sleep tends to coincide with when the brain produces large amount of a hormone known as 'prolactin'. This chemical has various effects, but can help generate a positive mood and so help to lessen the stresses and strains of everyday life. So, don't worry if you do wake up in the middle of the night. It may actually be good for you!

The young ones

Of course, there are two groups of people that will struggle to apply the rules of super-sleep: babies and young children. But worry not, help is at hand. Scientists have developed a series of techniques specifically aimed to helping these two tricky groups to enjoy the night. Here are seven of the best evidence-based tips to help you get through the night.

Routine

For years, many parenting experts have suggested that one of the simplest and most effective ways of improving children's sleep involves creating a bedtime routine. Research by Jodi Mindell from the Children's Hospital of Philadelphia shows that they are correct.[28] In Mindell's study, a group of parents were asked to carry out the same thirty-minute bedtime plan, which involved giving the child a bath, lightly massaging them, giving them a little cuddle, placing the child in their bed, and then turning out the lights. Within two weeks the children fell asleep faster, woke up less during the night, and were in a better mood in the morning. Other work suggests that the routine should avoid anything too stimulating (no television, video games, high-energy hide-and-seek, or caffeinated drinks). Some caregivers also find it especially helpful to create a 'bedtime chart' that shows each of the steps, setting a timer for twenty-five minutes and then rewarding the child if they finish before the timer goes off.

Crying it out

This option involves caregivers establishing a bedtime routine, putting the baby or child to bed, and then ignoring them no matter how much they scream and shout (or, as proponents of the technique refer to it, 'self-soothe'). Although seemingly tough, research shows that this approach often produces fast results and reduces severe crying within just a few nights.[29] However, some practitioners and parents have raised an eyebrow at the technique, suggesting that is overly harsh and that continued crying may cause psychological or physical harm to the child.*

Graduated extinction

This technique is often seen as the acceptable face of 'crying it

---

* The 'crying it out' and 'graduated extinction' exercises described here are designed to provide a general insight into these techniques and should not be used in place of a comprehensive guide. Neither 'crying it out' or 'graduated extinction' are usually recommended for children less than six months old.

out'. The caregiver starts off by staying with the child until they start to fall asleep, and then returns to their own bed or living room. If the child wakes up and starts crying, the caregiver waits a specified period of time before going to see the child and, even then, only offers minimal attention (excessive talking to or touching the child are frowned upon). After a few weeks the interval between the infant crying and the caregiver arriving is increased, typically starting at five minutes and increasing in five-minute blocks. Once again, research shows that this technique is effective.[30]

Retreat

Does your child struggle to sleep unless you are next to them? Then it is time to retreat. In this gentle technique, you start off by being next to the child and waiting until they are asleep. Then, a few days later, you move to the edge of the bed and again wait for them to nod off. Then, over time, you move further away from the bed, perhaps sitting in a nearby chair or sitting on the other side of the room. You have to resist any protests by the child and, over time, aim to have your child falling asleep without your having to be in the same room.

Fading

This two-part technique is designed for toddlers who are struggling to fall asleep until late in the evening, and so waking up late in the morning. First, the caregiver makes going to bed pleasurable so that the child associates bedtime with fun-time. This might involve playing a quiet game or two, reading a bedtime story, or creating a drawing. This routine is then instigated when the child feels sleepy, no matter how late this is in the evening. The caregiver then wakes up the child at a normal time in the morning (rather than waiting for them to wake up at their preferred time), and then moves the bedtime slightly earlier each night until the child is going to bed at a reasonable time.

Shut that door

Does your child keep wandering out of their room at night? If so, they need to learn how to fall asleep. This can involve giving them some kind of object to hug (such as a teddy bear or blanket) or getting them to make up a fun story in their heads. However, in the meantime, stop the wandering by immediately returning the child to their bedroom and shutting their door for one minute. If the child wanders again, increase the amount of time that the door is shut by a minute, up to a maximum of five minutes. After that, padlock their room (just kidding).

Lark-rise

Is your child getting out of bed too early? Try installing a night light with a timer, and tell them that they can only get out of bed when the light comes on.

Should you sleep in the same bed as a baby?

Many caregivers are tempted to sleep in the same bed as their babies to make it easier to soothe the infant and breastfeed them. But is this a good idea?

'Sudden infant death syndrome' (SIDS) occurs when infants below the age of one inexplicably die. Although the occurrence of SIDS in the US has dropped over 50 per cent since parents have been advised to put sleeping infants on their backs, it is still the third-leading cause of all infant deaths in America.

After analysing records from more than 1,000 SIDS deaths, Robert Carpenter from the London School of Hygiene & Tropical Medicine and his colleagues concluded that bed sharing should be avoided. Carpenter's findings showed that sleeping with an infant that was less than three months old resulted in a fivefold increase in the risk of SIDS, even among parents who did not use alcohol, drugs, or smoke cigarettes.[31] It seems that it's all too easy for sleep-deprived caregivers to inadvertently place a hand, or arm, over an infant's face or roll on top of them.

Health professionals have urged caregivers to avoid sleeping in the same bed as a child who is less than six months old. If caregivers do want to be close to a child throughout the night, it is suggested that they place the child in a crib, or cot, close to their own beds.

For centuries people have struggled to sleep. It doesn't have to be like this. Sixty years of research has resulted in a series of techniques that can boost both the quantity and quality of sleep in adults, babies, and children. All you need to do is learn the secrets of the super-sleepers, and then enjoy the best night's rest you have ever had.

ASSIGNMENT

The mouth and nose test

Please take a few moments to carry out this simple three-part test, and make a note of your results.

1) First, please close your mouth. Now shut your left nostril by gently pressing on the side of it. Keeping your mouth closed, take a deep breath through your right nostril. Now repeat the test, but this time, close your mouth and right nostril, and then take a deep breathe through your left nostril. Finally, keep your mouth closed, and take a deep breath through both nostrils. Did you feel like your nostrils were congested, and therefore breathing was difficult, during any of these exercises?

☐ Yes ☐ No

2) Please open your mouth and try to make a snoring sound. Now close your mouth and try to make the same sound. Are you able to make the same snoring sound with your mouth closed?

☐ Yes ☐ No

3) If you can make a snoring sound with your mouth closed, stick your tongue slightly out of your mouth and gently grip it with your teeth, ensuring that your lips are sealed around the sides of your tongue. Now try to make the snoring noise again. Is the sound of your snoring reduced?

☐ Yes ☐ No

Many thanks. We will come back to your results in the next lesson.

Lesson 4

ON SLEEPWALKING AND NIGHT TERRORS

Where we go for a stroll with some sleepwalkers, find out
if you can commit a murder in your sleep, and discover
the deadly downside of snoring.

Most nights the vast majority of people move through the various stages of sleep in a systematic and orderly fashion. However, once in a blue moon, some individuals don't follow this well-trodden path, and instead travel off-road to a weird and mysterious place. A place where the normal rules don't apply, they may encounter their inner demons, and they could even face death. In this lesson we are going to journey deep into the dark side of sleep, and find out what really does go bump in the night. During our time together you will discover just how wonderfully weird your brain can be, find out what happens when your unconscious mind suddenly hijacks your body, and learn something that might save your life. We begin with a spot of naked gardening.

A great perturbation in nature

In March 2005, Rebekah Armstrong was woken up by a strange noise emanating from her garden.[1] Noticing that her husband Ian was no longer in bed, Rebekah decided to go downstairs and find out what was going on. She wandered outside and discovered that her husband was standing in the middle of their garden completely naked. Even more surprisingly, Ian was busily cutting the grass with the couple's electric lawn mower, and had clearly been working away for quite some time because he had almost finished mowing the entire lawn. Perhaps most remarkable of all, he was sound asleep. Ian had gone walkies in his sleep several times before, but this was the first time that he had undertaken a spot of nocturnal naked gardening. Reluctant to disturb her semi-conscious husband, Rebekah unplugged the mower and went back to bed. A short while later, Ian sleepwalked his way back to the bedroom and lay down next to his wife. When Rebekah woke Ian up and explained what had happened, he initially refused to believe her, and only changed his mind when he saw that the soles of his feet were caked in soil and that the lawn was newly mowed.

Ian Armstrong is far from the only nocturnal wanderer to take a walk on the wild side. In June 2005, police and firefighters were called to a building site in southwest London because a passer-by had spotted a young girl lying on the arm of a 130-foot crane.[2] A firefighter carefully climbed up the crane, only to discover a fifteen-year-old girl fast asleep on the top of a large concrete counterweight. The eagle-eyed firefighter noticed that the girl had a mobile phone in her pocket, and so decided to quietly crawl up to the girl, gently remove the telephone, and call the girl's parents. The girl was eventually woken up with the help of her mother, and brought back down to earth via a hydraulic lift. Subsequent enquiries revealed that the girl lived near the building site, and had managed to climb up the crane and walk along its narrow arm while sound asleep.

Other sleepwalkers have displayed a similarly high level of disregard for their own safety. In August 2004, Basingstoke police found an abandoned and mangled car next to a busy road. The car appeared to have careered out of control, hit a lamp post, and ended up being embedded in the undergrowth. The driver was missing, but the police were able to trace the vehicle to a nearby pub. When officers arrived at the pub they found the car's owner wearing a blood-stained T-shirt and suffering from several facial cuts. The man was arrested and a breathalyser test revealed that he was three times above the legal limit. When the police interviewed the driver he denied having any memory of the incident, and insisted that the crash had happened while he was sleepwalking. The results of several medical tests, combined with a family history of sleepwalking, proved in court that the man was telling the truth, and he was found not guilty.

Some sleepwalkers prefer to spend the night eating rather than drinking. For example, fifty-five-year-old former chef Robert Wood from Fife. Wood regularly sleepwalks into his kitchen, and rustles up omelettes, stir-fries, and even the odd spaghetti bolognaise while completely unconscious.[3] Such 'sleep eating' is surprisingly common among sleepwalkers, and frequently results in them struggling to understand why they are mysteriously gaining weight.

Then there are the strange instances of 'sexsomnia', in which sleepwalkers masturbate, grope others, and even engage in full-blown sexual intercourse. In one recently reported case, a middle-aged Australian woman regularly left her house and had sex with strangers.[4] Like many sleepwalkers, the woman claimed to have no memory of these encounters, and the problem only came to light when she discovered condoms scattered around her house. The woman's name was not released to the media, presumably to both ensure her privacy and prevent a dramatic rise in local house prices.

Equally curious is 'sleep-drawing', in which a person displays only a modest amount of artistic ability when they are wide awake but creates amazing images while sound asleep. Last year, thirty-seven-year-old Londoner Lee Hadwin hit the headlines when he claimed to suffer from this rare condition.[5] Creatively nicknamed 'Vincent van Sloth' and 'Kipasso' by the media, Lee's nocturnal artwork includes detailed portraits of Marilyn Monroe and a series of circle-inspired abstract pieces. According to one newspaper report, some of Hadwin's pieces have been bought by celebrities for six-figure sums.[6]

Sleepwalking has also kept pace with technology. In 2009, for instance, the academic journal *Sleep Medicine* reported the first occasion of 'sleep email'[7] (described as 'zzz-mailing'), wherein a forty-four-year-old Spanish woman climbed out of bed, walked to the next room, turned on the computer, logged into her email account, and invited a friend around for dinner ('Come tomorrow and sort this hell hole out. Dinner and drinks, 4 p.m. Bring wine and caviar only.'). The woman was only aware of the message when her friend replied the following day.

Such extreme examples of sleepwalking are relatively rare. However, surveys show that a surprisingly large percentage of the population routinely carry out simple actions when they are asleep, including sitting up in bed, starting to get dressed, or wandering around the bedroom. Indeed, one recent poll revealed that almost 4 per cent of American adults – more than eight million people – have experienced at least one episode of sleep-walking in the past year.[8] These episodes usually take place a few hours after people have gone to bed, and the problem is especially prevalent among the young, with 15 per cent of children experiencing some form of nocturnal wandering.

While some people spend night after night walking the walk, others are equally busy talking the talk.

All talk, no action

Sleep scientist Arthur Arkin from City University of New York dedicated his life to listening to people as they slept, and his 640-page tome *Sleep Talking: Psychology and Psychophysiology* is widely seen as the definitive guide to night-time utterances. For years, Arkin carefully monitored sleep talkers in his laboratory and his observations shattered various widely held myths.

Many people think that sleep talking only involves the occasional mumble or slurred word, but Arkin discovered that many sleep talkers merrily chatted away for several minutes at a time. Indeed, in the 1960s, American songwriter and prolific sleep talker Dion McGregor decided to record his extensive night-time comments and then use them as the basis for an album. McGregor teamed up with Decca records and together they created and released *The Dream World Of Dion McGregor (He Talks In His Sleep)*. On this surreal album McGregor unconsciously mumbles his way through several deeply strange monologues, including those describing a starch-based diet, a tiny city populated by midget chickens and miniature tombstones, and an odd dinner party in which all of the chocolate eclairs have been laced with poison. Perhaps not surprisingly, the album was a financial disaster and has never been reissued. On the upside, McGregor's ramblings have attracted a small but loyal following over the years, and his record is now a highly sought-after collector's item.

There is also the thorny issue of whether sleep talking presents a reliable guide to people's innermost thoughts and feelings. You may have heard it said that dreamers never lie, but do sleep talkers openly mention secrets that they would never reveal when they are awake? It's tricky to know for sure. Over the years there have been many anecdotal reports of sleep talkers letting the cat out of the bag, including instances of them confessing to serious crimes and describing illicit sexual affairs. However, after monitoring hundreds of sleep talkers over several years, Arkin only came across two relatively trivial instances in which someone apparently let something slip. In one, a young male sleep talker muttered, 'Am I queer or something?' and then, when asked about the comment the following day, 'chuckled in apparent shame'. In the second instance, a sleep-talking man mentioned the initials 'C P W'. At the time, Arkin innocently thought that they stood for 'Central Park West'. However, when Arkin mentioned the incident to the man the next morning, the man's face turned red as he explained that the initials referred to the 'cosmic pussy whip' used by an overly dominant wife of a 'close friend'.

Finally, Arkin's work also revealed that it's possible to hold a conversation with someone who is sound asleep. After years of trial and error, Arkin discovered that the best way forward involves waiting until your friend or partner nods off, and produces some sort of initial mumble. This mumble is your cue to leap into action and ask a couple of straightforward questions that require a simple 'yes' or 'no' answer, such as 'Is your name Eric?' and 'Were you born in London?' Once you have managed to elicit a few responses, you can then move on to questions that are more interesting and require a longer reply, including, for example, those timeless classics, 'Have you ever found a goat overly attractive?' and 'What is your PIN?' After chatting with lots of unconscious volunteers, Arkin concluded that they tend to be an evasive and cagey bunch, with one dramatically ending a conversation by declaring 'I'm asleep'.

In both sleepwalking and sleep talking, people's unconscious minds seem to take control of their bodies. Most of the time the results are fairly harmless and are, at worst, the cause of much merriment in the morning. However, once in a while a far more sinister type of nocturnal phenomenon can rear its ugly head. Like sleepwalking and sleep talking, those experiencing this phenomenon have no idea what they are doing. But this time their actions can have the most horrific and tragic consequences.

Fright night

At 10 p.m. on 15 July 2008, retired couple Brian and Christine Thomas drove their camper van into the Welsh seaside village of Aberporth. The couple found a space in a car park at the edge of the village, turned out the lights, and climbed into bed. Just as they were drifting off to sleep, however, Brian and Christine were disturbed by a group of teenagers wheel-spinning their cars. Annoyed by the noise, Brian got up, drove the camper van to another nearby car park, and again tried to settle down for the night.

Just after midnight Brian thought he heard one of the teenagers breaking into the camper van. Brian lashed out, and in the ensuing struggle believed that he had strangled the intruder to death. It was then that Brian woke to discover that he had just experienced a terrifying episode in which his sleeping mind had become convinced that he was in danger. There was no teenage intruder. Instead, Brian had killed his wife.

Brian was examined by several sleep experts in the ten months prior to the trial. Leading that work was the man who had originally piqued my interest in the science of sleep, Dr Chris Idzikowski. Chris and his team visited Brian in jail, monitored him for several nights, and produced a report that would be used to help determine Brian's guilt or innocence.

What Chris discovered during his prison examinations was that Brian had regularly experienced similar types of night-time disturbances prior to his holiday. Indeed, these experiences had so disturbed his sleep that the couple often slept in separate rooms at home. Chris's technician Stevie Williams wired up Brian to an EEG system and spent the night sitting outside his cell, monitoring the data. A camera placed next to Brian's bed allowed Chris to see what was happening on the other side of the locked door. Midway through the night Chris saw Brian suddenly sit up and look around. A glance at the EEG data instantly revealed that Brian was still sound asleep. After other similar episodes, Chris concluded that Brian had probably been suffering from these sleep disturbances for years without receiving any formal treatment. This evidence, combined with a plethora of positive character statements, led to Brian being the first person in Britain to be found not guilty of murder on the grounds of a sleep disorder.

Brian's unconscious, but deadly, behaviour may well have been the result of him experiencing a deeply strange phenomenon known as a 'night terror'.

Almost everyone has had a scary dream or nightmare. However, night terrors are very different phenomena, and experienced by only around 6 per cent of children and 2 per cent of adults. In a typical episode, people become convinced that there is some kind of serious threat around them, and respond accordingly. The exact nature of the imagined threat varies from person to person, but often takes the form of huge spiders, an unwelcome intruder, a pack of evil dogs, or an all-powerful supernatural entity. Even though the person is still asleep, they may suddenly sit up in bed with their eyes wide open, scream and lash out. Thankfully, only a vanishingly small number of night terrors result in a lost life. Sleep terrors tend to take place during the first few hours of the night, with each episode lasting a few minutes and usually ending with the person simply going straight back to sleep. During these scary episodes people's heart rate often accelerates to more than 160 beats per minute, making it one of the most stressful events that the body can experience. As I mentioned in the introduction, I went through a period of having night terrors on a regular basis and can confirm that they are no fun.

Sleepwalking, sleep talking, and night terrors are all examples of a more general disorder known as 'parasomnia' (Latin for 'beyond sleep'). Researchers have long attempted to get to grips with this strange phenomenon, and their work has proved invaluable to the millions of people whose bodies come alive when they are sound asleep.

Probing parasomnia

In 1815 John William Polidori produced one of the first theses on sleepwalking while studying for a medical degree at Edinburgh University. In his essay, catchily entitled 'Dissertatio medica inauguralis, quaedam de morbo, oneirodynia dicto, complectens',* the nineteen-year-old Polidori argued that sleepwalking was a hypnotic state that was best cured with beatings, cold baths, and electricity.[9] Although the essay failed to make an impact on medical minds, Polidori's work played a key role in shaping modern-day literature. Acting as the personal physician to the poet Lord Byron, Polidori was part of a select group of writers who were challenged by Byron to each produce a ghost story. Drawing on his fascination with the night and sleepwalking,

---

* 'Inaugural medical dissertation concerning a disease called sleep walking'.

Polidori produced one of the first English-language vampire stories, *The Vampyre*. This groundbreaking work quickly caught the public's attention and has resulted in Polidori being seen as the creator of the vampire genre of fantasy fiction.

During the late 1830s, the chemist and creator of paraffin, Carl Ludwig von Reichenbach, also became fascinated with sleepwalking. After interviewing several sleepwalkers at length, von Reichenbach concluded that the strange phenomenon was brought about by the position of the moon and reflected the existence of an all-powerful energy called the 'Odic Force'. Ironically for the creator of paraffin, von Reichenbach's theory failed to set the scientific world on fire, and his beloved Odic Force has now been relegated to the file drawer labelled 'Nah'.

Next in line to try to unravel the mystery of parasomnia was the world-famous psychiatrist Sigmund Freud. In 1907, Freud addressed the Vienna Psychoanalytic Society and unveiled his new theory of sleepwalking. According to the father of psycho-analysis, sleepwalkers had a deep-seated sense of anxiety and were trying to make their way back to a place that had made them feel secure as a child. Over the years, Freudians have failed to produce any convincing evidence to support his somewhat strange theory, resulting in sleep experts around the world rejecting the idea on the grounds of wild speculation.

Fortunately, modern-day researchers have worked hard to slowly piece together a more scientific understanding of sleep-walking.

Some of this work has revealed that sleepwalking has a strong genetic component, with around 30 per cent of sleepwalkers having a relative who walks in their sleep, compared to just 17 per cent of non-sleepwalkers.[10] This genetic linkage sometimes results in several sleepwalkers wandering around the same household at night, including one case in which an entire French family of sleep-walkers woke up to discover that they were all sitting around their dining table.

Other work has explored what is going on in people's brains when they sleepwalk or experience a night terror. After monitoring people's brain activity throughout the night, researchers discovered that sleepwalkers are not acting out their dreams, and sleep terrors are not nightmares. As we discovered in Lesson 1, dreams and nightmares tend to take place during REM. In contrast, sleepwalking, night terrors, and sleep talking originate during deep sleep (associated with Stage 3 and Stage 4 of the sleep cycle).

Although the process is still not fully understood, many of these sleep scientists believe that sleepwalking and night terrors are the result of the brain struggling to move from deep sleep to wakefulness. During a normal sleep cycle, we've learnt you move from Stages 1 and 2 (light sleep), through Stages 3 and 4 (deep sleep), back to Stage 2, and then experience REM. This cycle takes about ninety minutes to complete, and is repeated several times throughout the night, and, as you move between the lighter stages of sleep and REM, you often experience micro-awakenings. These episodes of wakefulness only last about ten seconds or so, and you immediately go back to sleep afterwards. Because micro-awakenings are so brief, you are usually not aware of them.

Many sleep scientists believe that sleepwalking, night terrors and sleep talking are the result of your brain rapidly moving from deep sleep into a type of state normally associated with a micro-awakening. As a result, you end up in a strange zombie-like state in which you are not fully awake or completely asleep. Although one part of your brain is trapped in deep sleep and so almost completely unconscious, another part is functioning well enough for you to carry out relatively simple forms of behaviour on autopilot. As one sleep expert noted, sleepwalkers appear to be experiencing a 'third form' of consciousness that is most similar to instances in which you may have driven a familiar route home from work, but then realized that you have absolutely no memory of the journey.

These important discoveries explain why sleepwalking and night terrors are so different to dreaming and nightmares. For instance, let's examine the issue of movement. When you dream, almost all of your body is paralysed in order that you do not act out your dreams and hurt yourself. However, during sleepwalking and night terrors you are able to wander around and thrash out. Also, when researchers woke up dreamers and sleepwalkers and asked them to describe what was going through their heads, two very different types of reports emerged. Not surprisingly, those that had been in REM tended to describe a complex and colourful dream. In contrast, the reports from the sleepwalkers were far more like the basic and fragmented thoughts that occur outside of REM, and often just involved a very simple scenario, such as 'must take my dog out' or 'my house is on fire'.[11] Also, sleepwalking and night terrors usually take place within a few hours of a person going to bed because the initial part of the night is dominated by deep sleep. In contrast, most dreaming happens in the early morning because the latter part of sleep contains much more REM. Finally, children are more likely than adults to experience sleepwalking and night terrors, because they spend more time in deep sleep during the night.

Dream disorders

Not all parasomnia happens outside of REM. In fact, there are two disorders that can emerge when you dream.

First, some people experience an unpleasant phenomenon known as 'sleep paralysis'. This strange experience is associated with a highly regimented set of sensations that include waking up and feeling a crushing weight on your chest, sensing an evil presence, and seeing strange figures in the darkness.[12] These strange sensations are often attributed to the presence of demons, ghosts, or aliens. In fact, it's science at work. When you dream, your mind produces lots of strange images, and your brainstem blocks almost all bodily movement. However, once in a while you wake up from a dream in a somewhat confused state, with your brain becoming conscious but still experiencing the paralysis and weird imagery associated with dreaming. This terrifying combination causes you to feel as if you are being pinned to your bed and perhaps seeing some type of scary supernatural entity. Struggling to make sense of what is going on, you convince yourself that you are being pinned down by an all-powerful evil entity. However, this phenomenon can be overcome with the help of another simple technique. If you do happen to experience sleep paralysis, try to wiggle a finger or toe, or blink. Even the smallest of movements will help your brain lift the paralysis and bring you back to reality.

The second form of REM-based parasomnia is known as 'REM sleep behaviour disorder' and is, in many ways, the exact opposite of sleep paralysis. Rather than being paralysed when you are conscious, people move around when they are having a dream. During these episodes people essentially act out their dreams, often kicking, screaming, punching, and even jumping out of bed. The condition affects about one in two hundred people, with the vast majority of sufferers being men over the age of fifty. There is no relationship between the aggression displayed during the episodes and waking behaviour, with many people who are friendly during the day becoming aggressive when they experience the disorder. The problem is treatable with various drugs, and those suffering from the condition are often advised to remove potentially dangerous objects from the bedroom and place their mattress on the floor. The disorder appears to be due to a malfunction in the part of the brainstem that controls movement, and around a third of elderly people with the disorder develop Parkinson's disease within three years.

Other brain-based work has examined what causes normally non-violent people to sometimes thrash out during episodes of parasomnia. In 2000, neurologist Claudio Bassetti, from the University Hospital of Zurich in Switzerland, set out to scan a sleepwalker's brain when they went for a nocturnal stroll.[13] Bassetti invited a sixteen-year-old boy with a history of sleepwalking into his laboratory and monitored his brain activity throughout the night. The boy tended to sleepwalk several times each week and so the researchers thought they had a good chance of catching an episode. They were right. On the second night the boy stood up with his eyes open, moved around and uttered a few unintelligible words. The data revealed that the front parts of the boy's brain were inactive throughout the episode. This section of the brain plays a vital role in exercising self-control, and so explains why those experiencing parasomnia may become uncharacteristically aggressive and violent.

Much of the research into night terrors and sleepwalking can help those suffering from these unwanted nocturnal intrusions. Researchers have examined many of the factors that promote sleepwalking and night terrors, including fatigue, heat, and stress.

One study, for instance, conducted by Mathieu Pilon at the University of Montreal, looked at the effects of sleep deprivation and disturbed sleep.[14] Pilon arranged for people that had a history of sleepwalking to spend the night at his laboratory. The researchers ensured that the volunteers had either slept normally the night before or had been kept awake for the previous twenty-five hours. Pilon and his team then waited until the volunteers went into deep sleep and attempted to disturb them by sounding a buzzer. 30 per cent of the well-rested sleepwalkers, and an amazing 100 per cent of the sleep-deprived sleepwalkers, started to head off into the night. The study showed that the incidence of sleepwalking and night terrors are increased by sleep-deprivation and night-time disturbances, and emphasize the importance of sufferers having lots of rest, sleeping in rooms that are especially dark and quiet, and getting a larger bed if they are disturbed by their partner's night-time movements. This type of work has played a vital role in understanding parasomnia, but isn't without its dangers. In a similar study researchers managed to elicit night terrors almost at will, but were then physically assaulted by some of the volunteers during their episodes.[15]

Other work has also shown that many forms of parasomnia are brought on by heat, and so those wishing to avoid sleepwalking or a visit from an evil entity should turn down any heating and avoid thick duvets or covers. In addition, there is the issue of stress. In Shakespeare's most famous tragedy, Lady Macbeth sleepwalks because of her overwhelming guilt at the deaths of several people, including King Duncan and Lady Macduff. Recent research supports this idea, noting that sleepwalking is associated with depression and anxiety.[16] The relationship is far from trivial, with people suffering from depression being three times more likely to wander around at night. If you do feel anxious or down at night, consider carrying out a simple relaxation exercise, or even just listening to some quiet music, before you go to bed. If the symptoms persist or feel overwhelming, seek out the help of a qualified professional.

How to stop sleepwalking and night terrors in children

Lots of children sleepwalk and experience night terrors. If your child does suffer from these problems then help is at hand.*

First, spend a couple of weeks keeping a note of the time at which the episodes begin. Most children tend to sleepwalk, or experience a night terror, around the same time each night. Research suggests that these episodes occur about twenty minutes into a ninety-minute sleep cycle.

---

* The exercises described here are designed to provide a general insight into the sorts of techniques that are used by health professionals. If you believe that you or your child have a sleep-related problem, please consult a professional.

Second, having established the approximate start time, wake up your child about twenty minutes before the episodes usually begin. This procedure is designed to catch them at the start of a sleep cycle and so waking them up should not prove too difficult or disorientating. When you wake your child up, give them a sip of water and consider reading them a bedtime story to help get them off to sleep again.

Research shows that carrying out this procedure for ten consecutive nights eliminates night terrors in more than two-thirds of cases.

Sadly, the notion that sleepwalkers cannot injure themselves is a myth. In reality, sleepwalkers often come to some harm, with part of the problem being that they often fail to register pain or respond to obvious danger signals. In one instance, a seventeen-year-old German sleepwalker stepped out of a fourth-storey window, fell to the concrete pavement below, and continued to sleep despite having fractured an arm. In many ways, they got off lightly. Another sleepwalker who fell out of an open window almost ripped their arm off, and a third fractured their skull and spent several months in a coma.[17] And the downside of sleep-walking isn't just about the danger of falling. One sleepwalker was seen repeatedly punching a brick wall with their bleeding fists. According to witnesses, they had a determined smile on their face and seemed to be completely unaware of their bloody wounds. In rare instances, sleepwalking can even be fatal. A few years ago, an electrician from Wisconsin sleepwalked out of his home dressed in only his underwear and a fleece shirt. Unfortunately the outside temperature was a very cold – -27°C (-17°F) – causing him to die of hypothermia. If you are a regular sleepwalker, or live with one, lock away knives, fragile objects, and firearms. Also, consider fitting window locks on upper floors and, if you stay at a hotel, try to book a room on the ground floor.

Finally, Mark Pressman from the Jefferson Medical College in Philadelphia has answered the age-old question about whether it is safe to wake up a sleepwalker or let sleeping walkers lie.[18] Pressman surveyed thirty cases in which people had been violent in their sleep, examining whether they had been woken up by another person. The results show that waking up a sleepwalker is a very bad idea. In one case described by Pressman, a night shift supervisor fell asleep in his office. Around thirty minutes later another employee came into the office and attempted to wake up his colleague. Confused, the supervisor pulled out his gun and shot dead his colleague. In another instance, a sleepwalker grabbed a nearby knife and killed the man who was attempting to slowly bring him back to reality. If you do come across a sleepwalker, don't suddenly wake them up or even touch them. Instead, quietly repeat their name or a reassuring comment ('Everything is OK', 'You are safe').

Sleepwalking, sleep talking, and night terrors are strange. They are not, however, the only unusual phenomena that can come to haunt you during the night. In fact, there is a far more deadly problem that millions of people across the world face on a nightly basis. Unlike instances of parasomnia, this problem is not due to your unconscious mind hijacking your body. Instead, it has its roots in something that seems, at least on the face of it, to be far more innocent and down to earth. Your throat.

Falling down and running around

A surprisingly large percentage of people suffer from two deeply strange sleep-related disorders: 'narcolepsy' and 'restless legs syndrome'.

The term narcolepsy comes from the Greek word *'narke'*, meaning numbness, and *'lepsis'*, meaning attack. Sufferers suddenly fall asleep without any warning, with about three-quarters of them experiencing some form of sudden muscle weakness that may result in their jaw dropping open, their head slumping forward, and even their legs giving way. These attacks can happen several times a day, and can last from just a few minutes to up to an hour. Narcoleptics usually feel refreshed the moment that they wake up, but soon become sleepy again.

This curious phenomenon is caused by sufferers entering REM state without going through the other sleep stages, causing them to suddenly feel very tired and experience the paralysis associated with dreaming. Narcolepsy affects around one in every 2,000 people, with many sufferers finding their symptoms embarrassing. Medics advise those with narcolepsy to avoid heavy meals and alcohol, and to exercise regularly. Unfortunately, there is no cure for the condition, but various drugs can help control the symptoms.

There is also the seriously odd restless legs syndrome (or 'RLS' for short). Sufferers of RLS usually experience an overwhelming urge to move their legs when they lie in bed, and may also feel an unpleasant tingling or itching sensation in their legs throughout the night. Often they will be forced to climb out of bed and walk around until the pain slowly vanishes, only to find that it returns the moment they lie back down. RLS affects around 10 per cent of the population, often runs in families, and appears to be caused by an imbalance in a brain-based chemical called 'dopamine'. The good news is that the symptoms are often eased by people adopting a regular sleep schedule, carrying out moderate levels of exercise during the day (excessive levels of exercise can make the problem worse), cutting down on the amount of coffee and alcohol they consume, sleeping with a pillow between their legs, and carrying out some simple stretches just before they go to bed. In addition, several drug-based treatments have proved highly effective.

Ondine's Curse

Are you sitting comfortably? Then allow me to tell you a story from German folklore.

Once upon a time there was a beautiful water nymph called Ondine. Like all nymphs, Ondine was immortal, but stood to lose her immortality if she bore the child of a mortal man. One day, Ondine went out for a walk and met a handsome knight called Sir Lawrence. The two of them fell deeply in love and within a few weeks they were married. For the next few years all went swimmingly, with the two of them moving into a local castle and eventually having a child. However, in line with nymph-law, Ondine started to age after she gave birth and soon noticed that the shallow Sir Lawrence began to lose interest in her.

One afternoon Ondine was walking by the stables close to her castle and heard the all-too familiar snores of her husband. Curious, she went inside the stable and discovered her husband lying in the arms of another woman. Ondine woke up Sir Lawrence with a kick, pointed her magical nymph finger at him, and uttered a curse. She told her cheating husband that he could have his breath as long as he remained awake, but that it would be taken away from him the moment that he fell asleep. German folklore experts have not been able to discover the end of the story, but one assumes that it involves Ondine changing her Facebook status to 'single parent'.

The story of Ondine's Curse may sound like the stuff of fantasy, but for millions of people across the world it reflects a dreadful reality. To understand what is going on it's important to explore the science of snoring.

Engineers use the term 'kludge' to describe any clumsy and inelegant solution to a problem that works but is far from perfect. Your throat is an evolutionary kludge because it has to perform several different functions, including talking, breathing, and eating. Unfortunately, many of these functions are incompatible with one another. For breathing you need a strong straight pipe, while for talking you need a highly flexible tongue and upper airway that can create many different sounds. The kludge-like solution that is your throat is not entirely rigid, and unfortunately this can create a real problem during the night.

When you fall asleep, the muscles in your throat relax, causing your flexible airway to narrow. If this passage becomes too constricted, the airflow becomes turbulent rather than smooth and this, in turn, causes the sides of the throat to vibrate. When these vibrations become significant they create a rough sound that scientists refer to as 'snoring'. The narrower the airway, the greater the sound of snoring. Around 40 per cent of the population snore, with more men than women grunting their way through the night. Although snoring isn't necessarily a problem per se, it can disrupt concentration during the day. In one study, researchers tracked trainee doctors, looking at the relationship between snoring and their final grades. The results were remarkable, with 42 per cent of the snorers failing their exams compared to just 13 per cent of the non-snorers.[19] These nocturnal noises can also cause tension in a relationship, with snoring being the third leading cause of divorce in Britain.

Additionally, when the airway blockage becomes severe, the walls of the throat are pulled together and may actually stay shut. Because of this, oxygen levels in the bloodstream start to rapidly fall, and the person experiences an 'apnoea' (Latin for 'absence of breath'). The effects are far from trivial, with research showing that those suffering from apnoea can experience the type of low oxygen associated with going to the top of Mount Everest. In extreme cases, oxygen may not get into the system for around fifty seconds. As a result, the person's brain and body go into panic mode as they struggle for oxygen, causing a micro-awakening, during which the person gasps for air. The good news is that this gasp is usually successful, and helps restore oxygen levels. The bad news is that this deadly cycle will then repeat itself a few moments later and may re-occur between five and thirty times an hour.

Sleep apnoea is a serious problem. Left untreated it is associated with an increased prevalence of high blood pressure, heart attacks, strokes, obesity, type 2 diabetes, and cancer. Sleep apnoea also affects driving. In one study Spanish investigators found that those suffering from apnoea were six times more likely than others to be involved in a car accident.[20] Perhaps most shocking of all, those suffering from sleep apnoea are rarely aware of the problem. They may feel tired during the day, but have no idea why they feel this way and get used to living life with very low energy levels. As a result, it's important to be able to recognize the tell-tale symptoms of this night-time terror.

First of all, it is a myth that everyone who snores is experiencing apnoea. In fact, even the loudest of snores does not mean that a person is definitely suffering from the condition. It is more a question of a person's snoring or breathing being constantly interrupted by pauses and gasps during their sleep. Indeed, the most worrying sign is when snoring *stops*. In addition, people with sleep apnoea tend to fall asleep at work or while driving, are irritable, report getting headaches in the morning, are forgetful, feel anxious, and have a decreased interest in sex.

Obviously, not everyone who reports some of these symptoms has sleep apnoea, but people suffering from the condition report them far more than most. About 25 per cent of Americans have some form of sleep apnoea, with roughly 5 per cent experiencing severe forms of the condition (stopping breathing for at least ten seconds at least five times an hour). Amazingly, the vast majority of cases go undiagnosed. Sleep apnoea is almost twice as common in men as it is in women, and increases with age (roughly 10 per cent of adults over sixty-five suffer from the problem).

Good vibrations

The good news is that if you do snore there are various tried-and-tested techniques to help restore the sound of silence to the bedroom. Many of these techniques also help with sleep apnoea (although, of course, if you think that you might be suffering from sleep apnoea then you should seek specialist help immediately).

First, it is helpful to find out what kind of snorer you are. Just before the start of this lesson I asked you to carry out three simple exercises (see page 119). The results will help us discover whether the problem is with your nose, mouth, and/or tongue.

If you answered 'yes' to question one, then you have a blocked nose. If just one nostril appears blocked then this might be due to a physical abnormality, such as a twisted septum or polyps. You might find it helpful to try using adhesive nasal strips to pull your nostrils apart, and so help prevent them narrowing when you are asleep. If both sides of your nose appear blocked, and you don't have a cold, then you might be suffering from an allergy. If your nose only tends to become blocked at night, you might be sensitive to the type of allergens produced by the dust mites that tend to inhabit old pillows and mattresses. If you think this might be the case, try washing your bedding frequently at a temperature of at least 60°C, avoid putting old blankets on the bed, and place your pillows and – if possible – your duvets into plastic bags and then put them in your freezer for twenty-four hours at least once a month.

If you answered 'no' to question two, then there might be an issue with your mouth. If this is the case, then you probably sleep with your mouth open, and often wake up with a dry throat. You may benefit from a 'chin strip', which is essentially a strip of tape that runs under your chin and helps stop your mouth falling open while you sleep.

Finally, if you answered 'yes' to question three then your snoring might well be due to your tongue vibrating. Typically, you will have an unusual bite, wherein your lower teeth are behind your upper teeth when you close your mouth. If this is the case, you might want to think about using a 'mandibular advancement device'. This is a plastic gum shield that is designed to fit into your mouth, help push your jaw forward, and increase the space at the back of your throat.

It's quite possible that you might be one type of snorer, or be a combination of any two, or even all three.

In addition to these techniques, you might want to try losing weight, stopping smoking, drinking less alcohol, and trying to sleep with your head at a thirty-degree elevation by placing a foam wedge under your pillow. Also, it is important to avoid sleeping on your back, as your tongue and soft tissue in the throat are likely to fall backwards and obstruct your airway. If you find this tricky, try placing a tennis ball in a sock, and sew the sock to the back of your pyjama top (or whatever else you wear in bed). This will make lying on your back very uncomfortable, and so encourage you to turn on your side.

Many people also find it helpful to use a treatment referred to as 'continuous positive airway pressure'. Those using this approach place a mask over their mouth and nose when they go to bed, and then connect the mask to a small unit that uses mild air pressure to keep their airways open throughout the night. Strengthening the throat muscles also helps combat sleep apnoea. In 2009, researchers from the University of São Paulo Medical School in Brazil had patients suffering from sleep apnoea practise a series of tongue and throat exercises for thirty minutes each day for three months. The group experienced a 39 per cent drop in apnoea episodes after the treatment.[21] Similarly, in 2005 Milo Puhan from the University of Zurich and his colleagues gave sleep apnoea patients several didgeridoo lessons, and then asked them to practise the instrument for the following four months. Compared to a non-didgeridoo-playing control group, the patients experienced a significant reduction in apnoea and their partners reported fewer disturbances during the night.[22]

Sing, sing a song

Sleep scientists have come up with all sorts of exercises to strengthen snorers' throat muscles. Perhaps the most enjoyable techniques involve a series of specially designed singing lessons. In 2000, dramatherapist Alise Ojay from the University of Exeter brought together a group of twenty chronic snorers and asked them to spend three months completing a series of singing exercises for twenty minutes each day.[23] The researchers recorded the snorers for a week before and after the experiment, and discovered a significant drop in night-time noise.

On the basis of this work Ojay created a series of 'Singing for Snorers' CDs that encourage people to repeatedly perform specially designed vocal exercises to simple tunes. If you are not into the idea of inflicting your singing on your family and friends, you might want to try some of the following simple throat exercises. Remember to start slowly and gradually increase the number of exercises that you carry out over time.*

- Select your favourite vowel and repeat it out loud for two minutes, three times each day.

- Press the tip of your tongue behind your top front teeth, and then slide your tongue backwards and forwards along the roof of your mouth for three minutes each day.

---

* The exercises described here are designed to provide a general insight into the sorts of techniques that are used by health professionals. If you believe that you or your child has a sleep-related problem, please consult a professional.

- Purse your lips for thirty seconds.

- Open your mouth and move your jaw gently to the right. Hold this position for thirty seconds and then repeat the procedure on left side.

- Stand in front of a mirror, open your mouth and look at your uvula (the 'hanging ball' at the back of your throat). Contract the muscles at the back of your throat for a second or so, ensuring that the uvula moves up. Now relax. Repeat this process for thirty seconds.

At the start of this lesson I explained that we were about to explore the dark side of sleep. During our time together we have discovered that millions of people across the world suffer from strange sleep-related disorders every night of their lives. The problems cause them to wander around in a zombie-like state, come face-to-face with non-existent evil entities, and stop breathing several times an hour. Perhaps more importantly, we have seen that in each of these cases sleep scientists have worked hard to discover the causes of these terrifying experiences, and have developed techniques that can help people overcome them. There is now no reason for you, or your friends and family, to fear the night. Science has finally beaten the bogeyman.

ASSIGNMENT

Mid-term exam

We are now halfway through the course. During our time together we have uncovered the downside of sleep deprivation, examined what happens to your body and brain when you nod off, discovered the secrets of super-sleep, and found out how to confront the dark side of the night.

This seems like a good time to find out if you have been paying attention. I have come up with a short exam containing four multiple-choice questions. The exam will be carried out under test conditions, so please don't look back at the previous lessons, search for the answers on the web, or telephone a friend. Also, if you need to go to the toilet during the exam, put up your hand, and I will escort you from the room and have a chat about the importance of bladder control.

You have five minutes to answer the following questions.

1) **One or more of these facts about sleep deprivation is true. Please circle the fact, or facts, that you think are correct.**

a) Sleep deprivation causes around 100,000 road accidents each year in America.

b) Sleep deprivation can severely affect your willpower.

c) Sleep deprivation costs businesses an estimated $150 billion a year in lost productivity.

d) Sleep deprivation significantly increases the risk of diabetes, obesity, and death.

---

2) One or more of these techniques will help people fall asleep. Please circle the technique, or techniques, that you think are correct.

a) Create a fun fantasy world in your head.

b) Try not to fall asleep.

c) Force yourself to yawn.

d) Make a list of what you have to do the next day.

e) Ensure that there is a hint of lavender in the room.

3) One or more of these procedures should be followed when you place a young baby to bed. Please circle the procedure, or procedures, that you think are correct.

a) Lay the baby on their side.

b) Lay the baby on their back.

c) Let the baby sleep next to you.

d) Put the baby in their own bed or cot.

4) One or more of these techniques will help reduce snoring. Please circle the technique, or techniques, that you think are correct.

a) Stopping smoking.

b) Drinking less alcohol.

c) Drinking a glass of water before you go to bed.

d) Trying to sleep with your head at a thirty-degree elevation.

e) Avoiding sleeping on your back.

f) Avoiding watching films that feature pigs.

Many thanks. The answers are on the following page.

Answers

1) All of the statements are true, so give yourself a point for each one you selected.

Your score for this question: _____

2) All of the statements are true, so give yourself a point for each one you selected.

Your score for this question: _____

3) The correct answers are (b) and (d). To minimize the possibility of injury or sudden infant death syndrome, young babies should be put to sleep on their own, lying on their backs. Give yourself one point for each correct answer you selected.

Your score for this question: _____

4) The correct answers are (a), (b), (d), and (e). Drinking a glass of water before you go to bed and avoiding films featuring pigs will not prevent people snoring. Give yourself one point for each correct answer you selected.

Your score for this question: _____

Now add up your scores, and write your final score below.

Your total score: _____

Finally, use this table to discover your grade and feedback.

ASSIGNMENT

Your score	Grade	Feedback
14–15	A	Congratulations, you are a fantastic student, a cheat, or a liar.
11–13	B	Very good. You have obviously been paying attention and it has paid off.
8–10	C	Good. You have taken lots in, but there is still a little bit of room for improvement.
5–7	D	Not bad. You have the basics.
1–4	E	Oh well, we all have off days. It might be an idea to take a quick look at some of the lessons again.
0	Fail	Seriously? Wow. Here, have this conical-shaped hat to help you keep your head warm.

Many thanks. When people take these types of tests they often wonder if there are quick, effective, and effortless ways of boosting their brainpower. In fact, there are, and in the next lesson we are going to explore one of the most controversial and commonly misunderstood of these techniques: sleep learning.

Lesson 5

SLEEP LEARNING AND POWER NAPS

Where we bombard a town with secret messages, uncover
the surprising truth about sleep learning, and discover
the power of the six-minute nap.

At 10 p.m. on the 28 August 1942, Professor Lawrence LeShan crept into the boys' log cabin for one last time. LeShan made his way to the centre of the room and, in a firm but hushed tone, repeated the phrase 'My fingernails taste terribly bitter' 300 times. LeShan then tiptoed out, quietly closed the cabin door, and went to bed.

A psychologist, LeShan had just completed one of the strangest experiments in the history of sleep science.[1] His aim had been to discover if it were possible to get people to learn in their sleep. To find out, he visited a boys' summer camp in New York State, selected a group of children who were chronic nail biters, and split them into two groups. Each group was made to spend the summer in different log cabins, with one group receiving 'sleep learning' and the other acting as a control.

For the first part of the study LeShan had waited until the boys in the 'sleep learning' cabin were sound asleep. He then quietly entered the cabin, placed a phonograph in the middle of the room, and repeatedly played a recording of the scary 'fingernails' phrase. Two weeks into the experiment the phonograph broke, and so the dedicated professor visited the boys on a nightly basis and delivered the instructions himself.

This pioneering experiment was not without precedent. For

centuries Buddhist monks had allegedly whispered the contents of sacred books into the ears of sleeping trainee priests to help them absorb important scriptures.[2] The idea of sleep learning had also played an important role in science fiction. In 1911, the magazine *Modern Electrics* published a story by Hugo Gernsback entitled 'Ralph 124C 41+'. This curious work is set in 2660 and describes a series of inventions created by its central character (the afore-mentioned 'Ralph 124C 41+'), including a sleep-learning device known as the 'hypnobioscope'. According to the story, the hypno-bioscope was used across the world, and allowed children to learn information in their sleep and adults to have news reports fed directly into their brains throughout the night. Aldous Huxley envisioned a far more sinister use for the same type of device in his 1932 novel *Brave New World*. In Huxley's story, sleep learning is accidentally discovered when a young boy is able to recite a whole radio show after he falls asleep listening to the radio. The government then uses the technology to broadcast night-time messages that are designed to shape the population's opinions and morality.

After subjecting his unsuspecting nail biters to more than 16,000 suggestions, Professor Lawrence LeShan had a nurse assess the boys' fingernails. 40 per cent of those in the sleep-learning cabin had kicked the habit versus none in the control group. LeShan's findings suggested that the hypnobioscope may be more science fact than science fiction, and inspired decades of equally strange sleep-learning studies.

In the late 1950s, for instance, the head of Tulare County Jail in California decided to discover whether sleep learning could help reform prisoners. Small speakers were carefully placed under prisoners' pillows, and each night the sleeping inmates were subjected to hundreds of carefully chosen phrases emanating from a record player in the warden's office ('You shall grow in mind and spirit' and 'You will live without alcohol'). After a few years the prison authorities announced that their idiosyncratic approach to rehabilitation had been about '50 per cent effective', with newspaper reports describing how one prisoner said that the mere thought of drinking alcohol now made him feel sick and another assuring his captors that he could now go to sleep with a clear mind.

Fast forward to the 1960s, when Soviet researchers became worried that the Americans were winning the sleep-learning arms race, and consequently decided to conduct a mass participation experiment into the strange phenomenon.[3]

The small town of Dubna lies about 70 miles north of Moscow on the river Volga. At the time of the study about 20,000 people were living in the town and it was home to a major Soviet atomic research centre. Starting in December 1967, the Ukrainian Academy of Sciences required the good folk of Dubna to place radios in their bedroom and to ensure that they were tuned into a specified local station. For the next two months the residents were instructed to be in bed by 10.30 p.m., and keep their radios on throughout the night. At 10.40 p.m., the station broadcast fifteen minutes of light music to lull the town to sleep. Then, at exactly 11.05 p.m. and 6.30 a.m., the station broadcast a series of English language lessons. After endless nights of nocturnal bombardment, the scientists in charge of the project reported that the residents of Dubna had unconsciously learnt 1,000 English words and could now engage in simple conversations.

These striking findings suggested that sleep learning was a powerful tool that could be used to break bad habits, re-shape criminal minds, and teach a foreign language.

However, not everyone was impressed by the original research on prisoners, including William Emmons and Charles Simon from The Rand Corporation. Soon after the initial sleep learning studies were published, Emmons and Simon pointed out that the researchers carrying out the studies had frequently failed to check whether their volunteers were actually asleep when they were subjected to the night-time messages, and so argued that some of them may have consciously heard the messages.

To find out if this was the case, the duo conducted a now-classic study into sleep learning.[4] Emmons and Simon assembled a group of volunteers and locked them away in a sleep laboratory for the night. Each volunteer was connected to an EEG machine and the researchers carefully monitored their brainwaves. Whenever one of the volunteers definitely drifted asleep the researchers played an audiotape containing a list of ten words. This list was repeated as long as the EEG data showed that the volunteer was definitely asleep. In the morning each volunteer was shown a list of fifty words and asked to try to choose the ten words that had been repeatedly played. The volunteers were unable to reliably identify the words. Emmons and Simon's null results motivated other researchers to carry out similar studies. Time and again they discovered that any evidence of sleep learning vanished when they ensured that their volunteers were actually asleep. For years the scientific community had been growing increasingly sceptical about sleep learning, and this work proved to be the final nail in the coffin.

The idea of sleep learning remained unfashionable in scientific circles for more than thirty years. Then, in the 1990s, advances in sleep science inspired a new generation of researchers to take a fresh look at the topic.

Is there any reliable scientific evidence to suggest that sleep-learning CDs can help you master a foreign language in your sleep?

No.

How to develop a sleep-powered super memory

The German psychologist Hermann Ebbinghaus was one of the founding fathers of modern-day experimental psychology. Born in 1850, Ebbinghaus dedicated his life to psychology, and spent much of this time experimenting on himself. His most famous work involved staging several gruelling studies designed to uncover the mystery of human memory.[5]

Ebbinghaus wanted to figure out how his brain forgot information over time. Worried that his memory might be affected by past experience, the great German experimenter decided to create, and then try to remember, hundreds of non-existent 'words'. Each of the words created by Ebbinghaus contained three letters, had a 'consonant-vowel-consonant' structure, and was not already in existence ('CAT' was out, but 'CAX' and 'YAT' were fine). Before the start of each experiment he would draw up a list of about a hundred of these imaginary words, recite them again and again to the sound of a metronome, and then attempt to recall the list. Only after thousands of recitations, and near 100 per cent remembering, would the experiment begin.

Ebbinghaus then attempted to repeatedly remember his list

of words over the course of the next few hours, days, and weeks. After the first hour, Ebbinghaus discovered that he had forgotten around 10 per cent of the list. An hour after that and another 5 per cent had vanished. Three hours later and he was unable to recall yet another 2 per cent. And so it went on. After carefully plotting the number of remembered words at each point in time, Ebbinghaus discovered that his memory was decaying exponentially. This finding has stood the test of time, and his classic graph is still cited by modern-day memory researchers.

After carrying out several experiments into his decaying memory, Ebbinghaus noticed an anomaly in his graphs. Although the curves steadily dropped during the daytime, there was often very little loss of information over the course of the night. After much head scratching, researchers speculated that when you are asleep your brain rummages through your mind, throws away any facts and figures that it no longer needs, and stores away information that you have been trying to remember during the day. Several psychologists then conducted a series of studies to discover if this 'sleep is vital for learning' theory was correct.[6]

All of the studies were based on the same simple concept. Researchers first assembled some volunteers, and then randomly split them into two groups. One group was asked to remember a list of words in the morning and then tested in the evening. In contrast, the other group was given the list in the evening, and tested the following morning. As a result, each group had to try to remember the same set of words over the same time period, but the first group had to try to remember them during the day and the second group simply went to sleep at night. As predicted by the 'sleep is vital for learning' theory, those who spent most of the experiment sound asleep remembered far more words.

In short, whereas repeatedly playing phrases throughout the

night doesn't boost people's memories, sleep is essential for storing away information that you have encountered during the day. The take-home message is clear: do not skimp on sleep. When you are preparing for an important exam, or interview, you might be tempted to stay up late the night before trying to cram information into your head. Avoid the temptation. It's a terrible idea and you will be much better off getting an early night. Not only will you be more refreshed when you wake up, you will also be better able to remember what you learnt the day before.

The effect that a lack of sleep has on academic performance is far from trivial. A few years ago, Avi Sadeh from Tel Aviv University explored this notion.[7] Sadeh went into various schools, and randomly allocated fourth-graders and sixth-graders to one of two groups. Those in one group were instructed to go to bed thirty minutes earlier each night, while those in the other group were asked to stay up thirty minutes later than usual. Three days later researchers tested the children's performance on various educational attainments tests. The results revealed that the small amount of sleep loss was equivalent to the loss of two years of development, with sleepy sixth-graders performing like fourth-graders.

Unfortunately, millions of parents, children and students across the world fail to recognize the importance of sleep. In 2013, researchers from Boston College examined the educational attainment and sleeping habits of almost one million pupils from more than fifty countries. Their findings are staggering. Overall, nearly half of the children needed more sleep. America topped the 'sleep-deprived' league table, with an amazing 80 per cent of thirteen and fourteen-year-olds not getting enough sleep. Competing for the crown were pupils from New Zealand, Saudi Arabia, Australia, and England. Towards the bottom of the table, and therefore getting more than their fair share of sleep, were pupils from Portugal, the Czech Republic, Japan, and Malta. The project also revealed that the issue was especially severe in more affluent countries, with the researchers speculating that this was due to the excessive use of mobile phones and computers late at night. The problem even affected those pupils who were not sleep deprived, with many teachers having to dumb down their lessons to allow the large numbers of fatigued pupils to participate.

The situation becomes even worse when children hit their teenage years. As we discovered in Lesson 1, during adolescence people's circadian rhythms usually undergo a significant shift and so they feel especially sleepy in the morning. This phenomenon, combined with the relatively early start times of most schools and universities, has resulted in our higher education establishments being populated by legions of sleep-deprived teenagers. Several researchers have examined the effect of such deprivation, and the results make for grim reading. Psychologist Amy Wolfson from the College of the Holy Cross in Massachusetts surveyed more than 3,000 high-school students, and discovered that A- and B-grade students were going to bed about forty minutes earlier, and sleeping around twenty-five minutes longer, than those getting lower grades.[8]

However, the good news is that even getting a small amount of extra sleep has a surprisingly large impact. Several colleges have experimented with delaying their start times to fall in line with teenagers' natural internal clocks, and discovered that even the most modest of changes have a tremendous impact. For instance, one American high school in Minnesota changed its start time from 7.25 a.m. to 8.30 a.m. After the change, the average math and verbal SAT scores among the top 10 per cent of students had increased by 300 points. When another school in

Kentucky pushed back its start time by an hour, teenage car accidents dropped by nearly 20 per cent. A similar project in Britain cut persistent absenteeism by almost a third, and significantly boosted pupils' mathematics and English grades.[9]

Other work shows that the same effect may be able to help people who wish to forget, rather than remember, certain life experiences. In 2010, Kenichi Kuriyama, a neuroscientist from the National Center of Neurology and Psychiatry in Tokyo, carried out an intriguing study in which he showed a group of students several video clips filmed from the viewpoint of somebody driving along a street.[10] Some of the clips were entirely uneventful, while others contained a sudden and horrific car crash. Half of the students were then asked to stay up all night, while the others were allowed to sleep normally. Over the course of the next ten days, Kuriyama showed the students still photographs from the video clips, and asked to rate the degree to which the pictures made them feel anxious. The students that had been sleep deprived immediately after seeing the videos found the photographs significantly less stressful than other students, suggesting that the deprivation had prevented them from storing the negative images in their memory. This work suggests that those who are unfortunate enough to experience a traumatic event should perhaps try to avoid sleeping the night of the incident.

Encouraged by the overwhelming support for the role that sleep plays in remembering facts and figures, researchers began to explore whether the same idea also applied to other forms of learning.[11]

Why do babies sleep so much?

Most newborns sleep for around twelve hours a day, and researchers have long struggled to explain why they spend so much time in the land of nod.

In 2006, Rebecca Gómez and her colleagues from the University of Arizona speculated that when babies sleep they are storing information about how the world works, and that they sleep lots because there is a lot to learn.

Gómez decided to explore how naps affected babies' ability to learn a made-up language.[12] Her mythical language would consist entirely of nonsense words ('pel', 'hiftam', and 'jic') and follow some simple grammatical rules (for example, if a sentence started with 'pel', its third word would be 'jic'). She subjected a group of fifteen-month-old babies to hundreds of these sentences, ensuring that some of them heard the sentences before they took a nap, and the others heard them immediately after waking up. A few hours later, Gómez and her team placed speakers either side of their miniature volunteers, and played some new sentences to the babies. Some of these sentences followed the grammatical rules that had been established earlier and others didn't.

The results were remarkable. When babies hear something that they like, or that interests them, they turn their head towards it. In Gómez's study, the babies who had taken a nap after hearing the words were especially good at learning the abstract grammatical principles, and so spent longer listening to the sentences that conformed to these principles. This work suggests that babies do not sleep to simply get some rest, but are instead hard at work carefully analysing and cataloguing all of the experiences that they have just had when they are awake.

Let's get physical

Please try the following exercise using your dominant hand. Tap your thumb and first finger together. Thanks. Let's label that action 'T-1'. Now tap your thumb against your second finger, and let's label that 'T-2'. Finally, using the same notation, tap out the following pattern as quickly as you can:

T2 T3 T4 T2 T3 T1 T1

Now try to perform the sequence again. And again. All being well, your third attempt was faster, and more accurate, than your initial go. Researchers refer to this as the 'finger-tapping task', and use it to measure how quickly and accurately people can learn new physical skills. Following in the footsteps of those researchers interested in sleep and word lists, scientists had volunteers practise these finger-tapping patterns either first thing in the morning or late at night, and then tested them all twelve hours later.[13] Once again, the 'sleep is vital for learning' theory came up trumps. The volunteers that had been merrily tapping away in the morning showed the same speed, and accuracy, as when they were tested before they went to bed. However, the volunteers that had learnt the sequences late at night were much more accurate, and quicker, after they had slept. These improvements were far from trivial, with the group being 20 per cent faster and almost 40 per cent more accurate.

Subsequent research into this curious effect has revealed a few simple rules that can help those trying to learn to dance, master a musical instrument, drive a car, or play sport.

How to do anything

Do not get up too early in the morning

In one study, researchers asked volunteers to tap away during the day, and then monitored their brain activity through the night.[14] The longer the volunteers spent in Stage 2 (light sleep) during the two-hour period directly before awakening, the more their performance improved the following day. This finding, combined with the fact that we experience more light sleep in the early morning (between about 6 a.m. and 8 a.m. is vital), suggests that those trying to learn a physical skill shouldn't be jumping out of bed too early in the day.

Do not practise, rehearse, or train, directly before you go to bed

Research shows that the optimum time to go to sleep is about five hours after your training session.[15]

Practise different skills at least four hours apart

Studies show that practising a second skill immediately after the first stops any overnight improvement on the first skill.[16] Instead, make sure that you practise the skills at least four hours apart, or at least take a long nap between the two sessions.[17]

Try to obtain even more sleep than you require

In a series of studies, Stanford sleep researcher Cheri Mah monitored the effects of having high-level athletes sleep for ten hours each night for several weeks. The results have been startling, with Stanford elite swimmers shaving a tenth of a second off their average turn times and increasing their kick frequency by an average of five kicks,[18] tennis players significantly increasing the accuracy of their serves,[19] and basketball players improving their free throw shooting by almost 10 per cent.[20] While taking part in the studies, many of the athletes set new personal bests and broke long-standing records.

Take a nap

So far we have seen how a good night's sleep is essential to maximizing your ability to remember facts and figures, and learn new physical skills. However, when it comes to sharpening your mind with your eyes closed, this just the tip of the iceberg. In fact, there is something that you can do right now to help ensure that your brain is in tip-top shape. Even more remarkably, it will require just a few minutes of your time: take a nap.

Total recall

As people grow older they often struggle to remember names, dates, and everyday events. Many scientists believe that this memory loss may be a direct result of the elderly spending less time sleeping each night, and so not experiencing the quantity and quality of sleep necessary for learning.[21] To test this theory, some scientists have attempted to boost people's memories by using a strange technique known as 'transcranial direct current stimulation' (or 'tDCS' for short).

tDCS works by placing two electrodes onto someone's head and then connecting the electrodes to a small battery. The low current flows through the electrodes and the resulting electrical field effectively changes the sensitivity of the neurons in the area, and so alters the chances of them

---

responding to the signals they receive from other neurons. The equipment can be set to increase this sensitivity and so make the neurons more likely to respond, or reduce it and thus decrease activity in the neurons.

This technique has been around for a long time. In the early nineteenth century, researchers gained important physiological insights by applying electrical currents to cadavers and making their limbs move. In perhaps the most famous of these gruesome demonstrations, an Italian scientist named Giovanni Aldini travelled to London and 'reanimated' the body of a dead murderer.[22] The success of this work allegedly helped inspire Mary Shelley to write *Frankenstein*, and also encouraged Aldini to apply electricity to the skulls of melancholy patients in the hope of improving their mood. The technique was refined over the years and now forms one of the standard tools in neuroscience.

In 2004, Jan Born from the University of Lübeck in Germany set out to discover if tDCS could be used to boost sleep, and so improve people's memories for facts and figures.[23] Born asked volunteers to come to his sleep laboratory, remember a list of words and then head for bed. Although any type of sleep enhances memory, some work shows that learning facts and figures is especially associated with deep sleep (Stages 3 and 4 of the sleep cycle). Because of this, Born waited until each volunteer entered deep sleep and then applied tDCS to the front areas of their brains. This stimulation had the desired effect of encouraging the neurons in this area to produce even deeper levels of slow wave activity. The volunteers were then woken up and asked to remember the list of words. As a control, the researchers also arranged for the volunteers to carry out the same procedure on another night, but switched off the tDCS unit so that they didn't receive any electrical stimulation. Even though the volunteers didn't know which night their brains had been stimulated, the results revealed that they remembered more words when they had received the tDCS.

It is early days, but this type of technology could play an important role in helping people maintain their memories long into old age. In addition, if the theory is right, then other procedures that have been known to boost deep sleep, such as exercising more during the day or going to bed earlier than usual, may also help improve elderly people's memories.

The power of napping

Sleep learning does not require a full night of rest. In fact, even the shortest of naps can have a surprisingly big impact. In one Harvard study, for instance, volunteers were asked to try to memorize a list of words. Half of these volunteers were then allowed to take a twenty-minute nap while the others remained awake. When the researchers repeated the tests four hours later they found that the power nap had significantly boosted the volunteers' memories.[24] Exactly the same is true for those trying to learn new physical skills and abilities.[25]

Other research has shown that even a few minutes spent napping can make all of the difference. In 2008, for instance, German scientists from the University of Dusseldorf asked volunteers to memorize a list of words and then randomly allocated them to one of three groups.[26] The first group remained awake, the second slept for about forty minutes, and the third took a quick six-minute nap. When asked to recall the words, the Wide Awake Club did OK, the forty-minute sleepers did better, and those who nodded off for just six minutes came top of the class.

The effect even works on the youngest of minds. In 2013, psychologist Rebecca Spencer from the University of Massachusetts Amherst investigated whether a quick nap boosted the memories of children who were between three and five years old.[27] Spencer first asked the children to play a memory game in which they had to try to remember the location of various pictures in a grid. The children were then either encouraged to nap for just over an hour or remain wide awake for the same amount of time. Finally, the children's memory for the location of the pictures was tested the following day. The results provided more evidence for the power of napping, showing that the short nap boosted the children's memories by a remarkable 10 per cent.

Developing a super-powered memory is not the only psychological benefit to be gained through napping. Research by NASA revealed that pilots who take a twenty-five-minute nap in the cockpit – hopefully with a co-pilot taking over the controls – are subsequently 35 per cent more alert, and twice as focused, than their non-napping colleagues.[28] And, in 2009, sleep researcher Kimberly Cote from Brock University in Canada reviewed the vast amount of psychological work into napping, and concluded that even the shortest of snoozes causes significant improvements in people's mood, reaction time, and alertness.[29]

Psychologists are not the only ones to have become interested in the power of a catnap. Harvard University medical researcher Dimitrios Trichopoulos recently published the results of a six-year study into health and napping.[30] Trichopoulos and his team studied the lives of more than 20,000 adults aged between twenty and eighty. All of the participants were asked about their dietary habits, levels of physical exercise, and the extent to which they napped. Even after taking age and level of physical activity into account, those who took a thirty-minute siesta at least three times a week had a 37 per cent lower risk of heart-related death. Trichopoulos's finding may explain why coronary disease is very low in cultures that encourage daily siestas.

Additional work by Mohammad Zaregarizi from Liverpool John Moores University suggested that these nap-related health benefits may be due to a lowering of blood pressure.[31] Zaregarizi monitored the blood pressure of people as they spent up to an hour either napping, standing up wide awake, or lying down without sleeping. Only those napping showed a significant drop in blood pressure. Zaregarizi's results also revealed that the napping volunteers experienced the biggest drop in blood pressure in the time between them lying down and falling asleep, showing that just the expectation of a nap is good for your body and that even a shortest of snoozes will benefit your health.

Napping is often seen as a form of laziness. Nothing could be further from the truth. Hundreds of experiments have demonstrated the enormous benefits associated with even the shortest of sleeps, and so it is vital that you make napping part of your daily routine. Putting your head down for just a few minutes each day will help you develop a better memory, be more alert, increase your reaction time, and boost your productivity. Perhaps most important of all, it may even save your life.

Inspired by the psychological and physical benefits of napping, researchers have explored many aspects relating to the science of snoozing. The following FAQ will help you enjoy the perfect power nap.

Where should I nap?

Find a dark, quiet, place where you won't be disturbed by telephones, emails, doorbells, pets, or other people. Your body temperature will drop when you fall asleep, so choose a warm room or wrap yourself in a light blanket. If your chosen room is not especially dark or quiet, try wearing an eye mask and earplugs. Some people (especially those of an introverted disposition) also find it helpful to listen to some sort of constant background sound, such as a relaxation tape, a soporific podcast, or monotonous sports commentary.

What if I am at the office, and my only option is to lean forwards onto a desk?

Excellent question. In 2012, Dayong Zhao and his colleagues from the Southwest University in China, attached a group of experienced nappers to an EEG machine, and then measured their attentiveness by having them listen to a string of musical notes and spot the occasional one that had a different pitch.[32] The volunteers were then randomly assigned to one of three groups. One of the groups enjoyed a twenty-minute nap lying down, another was given a pillow and asked to nap by leaning forwards onto a desk, and the final group spent the same amount of time sitting quietly. Afterwards, all of the volunteers' attentiveness was measured a second time. Compared to those that had just been sitting quietly, both the leaning and lying nappers were in a better mood, felt less sleepy, and were more attentive. Interestingly, those that had been lying down and leaning forwards rated their naps as equally beneficial. However, the EEG and attentiveness data told a different story, showing that those who had been lying down had obtained a greater amount of deep sleep and woke up significantly more alert. Lie down if you can, but even grabbing a quick nap while leaning forwards has clear benefits. Oh, and if your boss isn't happy about you catching a quick nap at work, show them the vast amount of research demonstrating that getting your head down for a few minutes will make you far more productive, successful and creative than your non-napping colleagues.

How long should I nap for?

Lots of research has examined how long you should spend in the land of nod. There is no magic rule about the perfect power nap as each nap length is associated with different benefits. Here is a quick guide.

The Micro Nap (less than 5 minutes)
This nap doesn't produce much in the way of psychological or physical benefits, but if you are very tired it will help you shed some of the sleepiness.[33]

The Short Nap (10–20 minutes)
During this type of nap your brain will mostly be in light sleep (Stages 1 and 2), but may experience a small amount of deep sleep (Stages 3 and 4) towards the end of the session. The large amount of light sleep will result in you feeling much more alert and focused when you wake up. Evidence also suggests that this type of nap will help improve your 'muscle memory' for new skills and abilities, and will yield several health-related benefits, including a lowering of blood pressure.

The Long Nap (20–60 minutes)
During this longer nap your brain will spend lots of time in both light and deep sleep. In addition to having all of the benefits

associated with The Short Nap, this will promote your ability to learn facts and figures. In addition, your brain will also start to release growth hormones, resulting in you feeling more energized when you wake up. However, you might emerge feeling groggy because you are waking up from a deeper sleep, but this should wear off after about thirty minutes.

The Full Nap (60–90 minutes)

During this nap your brain will complete a single sleep cycle, experiencing light sleep, deep sleep, and the REM state. These naps have the benefits associated with The Long Nap and, in addition, the REM state helps enhance your creative thinking and ability to grasp abstract concepts (more about this in Lesson 7). In addition, you shouldn't feel groggy when you wake up because you will be coming round after a REM period.

When should I nap?

Most people feel the need to nap in the mid-afternoon because this is when a dip in their circadian rhythm makes them feel especially tired and lethargic. However, sleep expert Sara Mednick, from the University of California, has examined how the time that you wake up in the morning influences when you should nap in the afternoon. In her book *Take a Nap! Change your Life*, Mednick argues that the perfect nap should be ninety minutes long, and contain the same percentage of light sleep, deep sleep, and REM that we experience during the night. According to her research, this unique combination is most likely to provide the maximum level of psychological and physical benefits. If you want to give Mednick's theory a try, use the following table to discover when you should take a nap.

Time you woke up in the morning	Perfect nap time
6.00 a.m.	1.30 p.m.
6.30 a.m.	1.45 p.m.
7.00 a.m.	2.00 p.m.
7.30 a.m.	2.15 p.m.
8.00 a.m.	2.30 p.m.
8.30 a.m.	2.45 p.m.
9.00 a.m.	3.00 p.m.

What if I feel guilty about napping?

It's vital that you get rid of any lingering doubts about whether napping is a good use of your time. Instead, remind yourself that naps can make you more alert, improve your reaction time, help you to become more creative, reduce accidents, and put you into a better mood. In fact, you should start to feel guilty if you are *not* taking a nap during the day.

What if I worry about oversleeping?

To ensure that you don't snooze too long, set an alarm. Oh, and don't worry if you don't fall asleep. Research shows that even just lying down with the intention of napping is enough to cause a healthy reduction in your blood pressure.

Do you have one other great final tip?

If you need to feel wide awake directly after having a short nap, drink a cup of coffee or other caffeinated drink just before dozing off. The caffeine will start to work its magic about twenty-five minutes later – just as you are waking up.

Suite dreams

In 2010, Google purchased several high-tech 'nap pods' for employees at their Googleplex headquarters in California. These futuristic devices consist of a specially designed reclining chair and a large spherical hood about a meter in diameter. Those wishing to take a quick nap lie down on the chair, and so are placed into a position that promotes blood circulation and reduces pressure on the lower back. The spherical hood is then placed around their head, and for twenty minutes or so they listen to low-frequency rhythms that aid relaxation and help to eliminate surrounding noise. When it is time to return to the real world, the person is gently woken using a combination of lights and vibration. The benefits of the pods quickly became apparent, with Google reporting that even the shortest of naps re-invigorated employees and made them ready to face the remainder of the day.

Google are not the only organization to wake up to the power of sleep. However, creating the perfect space for a quick snooze doesn't need to involve buying a high-tech nap pod. In fact, it can be done quickly and economically. A few years ago I took over a large room at my university, flooded it with a strange shade of green light (proven to enhance the brain's production of a 'feel-good' neurotransmitter called dopamine), and lit the ceiling to resemble a beautiful blue sky. Volunteers were asked to lie down on a soft yoga mat, rest their head on a lavender-scented pillow, and listen to a specially commissioned soothing soundtrack. The room acted like the ultimate sleeping potion, with all of the volunteers finding it remarkably easy to spend a few minutes napping. After about twenty minutes enjoying a light sleep they woke feeling especially alert and refreshed.

Also, there is the question of 'rocking'. Sophie Schwartz, from the University of Geneva, asked volunteers to take forty-five-minute afternoon naps on an 'experimental hammock' that either remained stationary or rocked gently from side to side.[34] Throughout each session the volunteers had their brain activity monitored. Even a small amount of gentle rocking dramatically increased the amount of deep sleep experienced by the volunteers, suggesting napping in a hammock or a rocking chair is a scientifically sound way of boosting your brainpower.

We are almost at the end of our journey into sleep learning. There is, however, one final twist in the tale. At the very start of the lesson I described how researchers had whispered into sleeper's ears in an attempt to stop them biting their nails, suppress their criminal tendencies, and learn a foreign language. The work failed because it was based on an incorrect understanding of the sleeping mind. However, several modern-day sleep scientists have recently followed in the footsteps of these pioneering researchers, and re-visited the idea of presenting people with information when they are sound asleep. These scientists have not, however, been creeping into log cabins or bombarding Russian towns with secret radio messages. Instead, they have been encouraging them to smell the roses, and to get in touch with their inner Jimi Hendrix.

Neuro-napping

In 2007, Jan Born, from the University of Lübeck in Germany, announced the results of a strange study into sleep learning. At the start of the experiment, Born had laid several picture post-cards on a table in his laboratory, and sprayed a subtle rose scent into the air.[35] Volunteers went into the room one at a time, tried to memorize the location of the postcards, and then spent the night in Born's sleep laboratory. Each volunteer was randomly assigned to one of two groups. Whenever those in one of the groups entered deep sleep (Stage 3 or 4), Born wafted the rose scent under their nose. Those in the other group had a scent-free night. The following day, each of the volunteers were asked to remember the location of the postcards. Born speculated that the night-time scent would remind the volunteers' sleeping brains about the postcards, and so would enhance their performance in the memory test the next morning. The results, published in the world's top academic journal, *Science*, showed that he was right – the volunteers who had smelt the roses in their sleep remembered the location of far more postcards.

This is not the only modern-day study to demonstrate that it is indeed possible to learn in your sleep. Ken Paller is a psychol-ogist from Northwestern University in America, who wanted to discover if people became better musicians if they were subjected to music in their sleep.[36] Using an arrangement reminiscent of the computer game Guitar Hero, Paller first asked volunteers to learn two new tunes by using their left hands to tap the songs out on a computer keyboard. These volunteers were then connected to an EEG system, asked to put on a pair of headphones, and encouraged to take a ninety-minute nap. When the EEG output indicated that the snoozers were in deep sleep, the researchers surreptitiously played one of the songs through the headphones. The volunteers were then woken and asked to try to tap out both of the tunes again. The results revealed that the volunteers were better at playing both songs post-siesta, but that they were especially good at playing the song that they had 'heard' while sound asleep. This work lies at the heart of a new form of sleep learning that I have labelled 'neuro-napping' (see below).

A little night music

This procedure is based on the latest research into sleep learning, and involves a new technique that I refer to as 'neuro-napping'.

- The procedure lasts about an hour and is best carried out in the early afternoon.

- Choose a piece of music that doesn't have any strong associations for you, that you find pleasant to listen to, and is relatively quiet. Any piece of instrumental, non-jarring, highly repetitive, 'background music' will usually work well.

- Play the music quietly in the background while you spend around twenty minutes preparing for an exam or interview, rehearsing for a presentation or play, learning a new language, practising a new dancing or juggling routine, training for a sport, or trying to find a creative solution to a problem.

- Turn off the music for ten minutes, and take a break.

- Now turn on the same piece of music again, lie down, close your eyes, and take a nap for about thirty minutes. If you struggle to nap, try following the guidelines described earlier in this lesson.

The music will encourage your brain to continue working away on the exercise as you sleep, and when you wake up you will discover a significant improvement in your mind and body.

Scientists' work has revealed that we all sleep learn every night of our lives. Although it may feel like your brain switches off when you go to bed or take a nap, research shows that your unconscious mind is busy shifting through your thoughts, deleting those that you no longer need, and replacing them with new information. Perhaps most importantly of all, when you understand how this process works, you can use it to boost your memory, master a musical instrument, become better at sport, and be more creative. Sleep learning is not the stuff of science fiction. It's just yet another example of your sleeping brain in action.

ASSIGNMENT

The Night School Dream Diary

We are about to begin our journey into the heady world of dreaming. To get the most out of the next few lessons it would be handy if you knew more about the events that take place in your dreams. The following tried-and-tested procedure has been developed by sleep scientists to help people remember their dreams.

1) Find a notebook. Some people like to use an expensive leather-bound journal, and others go with something far more economical.

2) Place your notebook, and a pen or pencil, on your bedside table.

3) Just before you go to bed, write this list of words down the left-hand side of a new page in your notebook, ensuring that there are equally sized gaps between the words.

Cast:

Location:

Plot:

Period:

Emotion:

Other:

4) As you are falling asleep, tell yourself that you will remember your dreams. Repeat this instruction in your head three times.

5) When you wake up after a dream, or first thing in the morning, do not open your eyes or move around. Instead, lie with your eyes shut and allow the fragments of your dream to drift together. If you are struggling to wake up after a dream, try drinking lots of water before you go to bed to encourage night-time awakenings. If all else fails, set your alarm clock to sound about six hours after you have gone to bed.

6) As you lie in bed, try to remember more about your dream by treating it like a play. Ask yourself the following type of questions.

Cast: Who was involved in the dream? Did you know the people or were they strangers? How were they dressed? Did the dream involve any animals or non-human figures?

Location: Where did the dream take place? Was it indoors or outdoors? Did you recognize the location?

Plot: What happened in the dream? Was there a narrative thread running through it, or was it more random?

Period: What was the time frame? Was the dream set in the past, present, or future? Were you the same age as now, or were you younger or older?

Emotion: How did you feel in the dream? Most dreams elicit negative emotions, so don't worry if the dream did not make you feel especially good. However, try to identify exactly how you felt.

Other: Was the dream in colour, or black and white? What do you think caused the dream? Does it seem to contain any obvious symbolism that relates to what is happening in your life?

7) Now open your eyes and write your answers in your notebook. Perhaps fill the page following with any drawings or diagrams that help illustrate your dream.

Please try to keep your dream diary each night for about two weeks. Don't worry if you struggle to remember your dreams during the first few nights. Dream recall takes practice and patience. Persevere, and by the end of the two weeks you will find that you are able to produce a surprisingly detailed description of your night-time fantasies.

Lesson 6

WELCOME TO DREAMLAND

Where we spend time at an ancient Greek sleep sanctuary, discover if you have the gift of prophesy, and uncover the secret symbolism in your dreams.

Each night, millions of people journey to an imaginary land and encounter a colourful cast of curious characters, strange animals, and supernatural entities. Unfortunately, trying to find out exactly what goes on in people's dreams is extremely problematic. The vast majority of folk struggle to remember their night-time adventures and, at best, usually produce reports that feel patchy and incomplete. As the modern-day sleep researcher Rosalind Cartwright once noted, attempting to investigate dreams using these fragmented reports is much like trying to figure out how jokes work when you can only hear the punch lines.[1] Because of this, the true nature of these weird experiences remained a mystery for centuries. Then, in the early 1950s, everything changed.

In Lesson 1, I described how Eugene Aserinsky transformed the science of dreaming overnight. Aserinsky's work centred on a strange stage of sleep that he labelled rapid eye movement (REM). When people enter this stage their brains become highly active, their eyes dart from side to side, their sexual organs get all frisky, and their bodies become almost completely paralysed. Aserinsky discovered that if someone is woken up directly after they have experienced REM, they almost always produce a long and detailed description of a dream. This simple discovery provided sleep researchers with a direct route into the dreaming mind. To return to Cartwright's comedy analogy, sleep scientists were no longer limited to working with punch lines, but instead had unlimited access to complete jokes and entire routines. They quickly set about exploring this brave new world, and their work has uncovered the surprising truth about the strange events that take place in your mind each night.

We are going to follow in the footsteps of the intrepid sleep scientists who used Aserinsky's discovery to test two of the oldest ideas about why we dream. Along the way, we are going to see whether your dreams can predict the future, meet one of the world's most influential thinkers, and find out if you have a dirty mind.

Let's begin with a spot of time travel.

A gift from the gods

Many people living in the ancient world assumed that dreams were supernatural in nature. For instance, the good folk of ancient Mesopotamia believed that when a person fell asleep, their soul left their body and went walkies, and dreams were the result of people seeing the world through the eyes of their wandering soul. This blurring of fact and fiction occasionally caused more than its fair share of confusion, with, for example, entire communities being thrown into panic when several people dreamt about seeing enemy forces approaching. In contrast, the ancient Babylonians were equally convinced that dreams were a gift from the gods, with pleasant dreams coming from benign deities and nightmares being sent by demons. These approaches to dreaming, although fascinating, pale into insignificance when compared to the elaborate ideas and schemes developed in ancient Greece.

The ancient Greeks believed that certain dreams contained god-given messages that were designed to help people make important life decisions. Around 200 BC, dodgy priests started to exploit this belief by charging wealthy citizens to attend special 'sleep sanctuaries'. These sanctuaries were constructed on sacred sites, and claimed to help people receive advice from their favourite deity.

To find out what these sleep sanctuaries were all about, let's travel back in time and imagine that you are a well-heeled merchant in ancient Greece. Nice tunic. Unfortunately, all is not well in your life because you think that your wife no longer loves you and, worse still, that she might be having an affair with the teacher next door. After months of indecision you finally decide to grab the bull by the horns and ask for some the advice from the goddess of love, Aphrodite. Eager to get the attention of this busy goddess, you contact your local sleep sanctuary and ask for an appointment. Luckily, the sanctuary has just had a cancellation and so they are able to fit you in quite quickly (apparently the god of time, Chronos, had to pull out of a session because he had double-booked). You are instructed to spend the forty-eight hours before your visit completely abstaining from both food and sex. Given your current marital situation, only one of these proves challenging.

The big day has now arrived. Hungry and horny, you make your way to the sanctuary, meet a priest, and hand over your hard-earned drachma. The priest asks you to sacrifice a goat, and then leads you through a series of catacombs until you arrive at a sacred room. Struggling to see in the darkness, you are just able to make out a bed-sized slab of stone. As your eyes slowly adjust to the low light, you see that the slab is covered with the skin from your recent sacrifice and that the floor is a seething mass of snakes. The priest explains that these animals symbolize transformation, and during your time at the sanctuary he hopes you will change your life in the same way that a snake sheds its skin (this symbolism has stood the test of time, with the logos of many modern-day medical organizations containing snake-like imagery).

The priest then asks you to hold a small statuette of Aphrodite in your clenched fist, and wraps a band of cloth around your hand to ensure that this sacred object remains close to you throughout the night. Finally, you are asked to carefully pick your way through the snakes, lie on the stone bed, and settle down for the night. Accompanied by the gentle whiff of freshly slaughtered goat and the soothing sound of slithering snakes, you try your best to nod off.

Throughout the night, the priest quietly comes over to the bed to ensure that all is well, and encourages you to sip water from his sacred bowl. As the night wears on you experience several bizarre dreams before apparently hearing Aphrodite whisper in your ear. In the morning you are woken up by the priest, and asked to describe the dreams and messages that came to you during the night. The priest then uses this information to figure out your best course of action. On this occasion it's bad news, with your dreams and messages strongly suggesting that the next-door teacher and your wife are not spending their time together studying moral philosophy. As dawn breaks, the priest hands over the bloodied goatskin as a memento of your time at the sanctuary, thanks you for coming, and sends you on your way.

Were the strange phenomena that you experienced at the sanctuary evidence of the supernatural power of dreams? Probably not. Many modern-day sleep scientists have argued that the rituals associated with the sanctuaries, including the fasting, the goat slaying, the snakes, and the tying of the hand into a fist, would have all helped create especially unusual and vivid dreams. Others have gone even further, accusing the priests of engaging in various forms of jiggery-pokery, including lacing the 'sacred' water with opium and producing the god-like voices via ventriloquism (thus explaining why many of the allegedly divine messages contained the phrases, 'gottle of geer' and 'I'm not coming out of the box').

Despite the deeply dodgy nature of the ancient Greek sleep sanctuaries, the association between dreaming and the supernatural has stood the test of time. By far the most popular paranormal take on the topic suggests that dreams may provide a brief peek into the future. Although the notion may sound a tad far-fetched to those of a more sceptical bent, the whole of human history is littered with evidence in favour of prophetic dreaming. To take just a few examples: the night before the battle of Waterloo, Napoleon foresaw the defeat of his forces; Abraham Lincoln dreamt about being killed two weeks before he was assassinated; and, in the Bible, Joseph uses the Pharaoh's dreams accurately to predict that Egypt will experience seven years of famine.

The majority of modern-day minds also buy into the prophetic power of dreams. In 2009, a team of researchers from Carnegie Mellon University asked hundreds of commuters in Boston to imagine that they were about to take a flight, and to rate how anxious they felt about the trip.[2] Before making their ratings, each of the volunteers was asked to think about one of four scenarios. One group were told to imagine that the government had just announced that there was a high likelihood of a terrorist attack, another group were instructed to think about their plane crashing, the third group were told that another plane had just crashed on the same route, and the final group were instructed to imagine that they had dreamt about their plane crashing the previous night. You might think the volunteers would feel most anxious after hearing about the government warning, or that another plane on the same route had just nose-dived into the ground. Amazingly, the volunteers that imagined having a plane-rated bad dream were the most likely to tick the 'high anxiety' box.

Is there anything to this belief? Is it really possible for your dreaming mind to take a peek into the future? If so, what does that tell us about the nature of dreams? And, if not, why do so many people have these weird experiences? To find out the answers to these questions and more, we are going to join a group of maver-ick researchers as they stage some of the strangest studies in the entire history of sleep science.

The Prince of Percipients

During the early 1960s, several kaftan-wearing, pot-smoking, paradigm-shifting, self-development gurus became convinced that dreams were a vast powerhouse of paranormal energy. Learn to tap into this strange state of consciousness, ran the argument, and you will be able to develop amazing magical powers and enjoy the gift of prophesy. This mystical take on dreaming chimed with the then-popular notions of unlimited human potential and psychedelic exploration, and so quickly spread from one tie-dyed commune to another. Soon, every self-respecting hippy was hang-ing a dreamcatcher on their window, lying down and tuning out, and trying to unlock the magic in their sleeping minds. Energized by this sudden interest in the supposedly paranormal power of dreaming, a handful of researchers working at the fringes of sleep science decided to discover whether there was anything to these bizarre claims.

Much of this research was conducted by a thirty-one-year-old New Yorker named Montague Ullman.[3] At the time, Ullman was training to become a fully fledged Freudian analyst and had noticed that many of his patients' dreams seemed to predict the future or involve some form of telepathic communication. Intrigued, the energetic Ullman teamed up with a parapsychologist named Stanley Krippner, and together they established a 'dream laboratory' at the Maimonides Medical Center in New York. Ullman and Krippner spent the next ten years carrying out a series of studies examining the possibility of dream telepathy. In a typical study, one volunteer (referred to as the 'percipient') would be connected to an EEG machine and spend the night sleeping in the laboratory. Meanwhile, a second person (the 'agent') would be placed into a nearby room and given a randomly selected art print (the 'target'). Whenever the EEG data suggested that the percipient was dreaming, the agent would be instructed to focus their attention on the art print. Then, a few moments after the EEG data showed that the dream had finished, the percipient would be woken up, and asked to describe their dream in as much detail as possible. Finally, an experimenter and the percipient would rate the degree of similarity between the dreams and a small collection of art prints, including the 'target' print. To avoid possible bias during this rating procedure, neither the experimenter nor the percipient knew which of the art prints had served as the target. If the target print was judged to be similar to the dream, the session was designated a hit. Otherwise, it was seen as a miss. Many of the experiments obtained statistically significant results, and seemed to suggest that something spooky was going on. Thrilled with these findings, Ullman and Krippner eventually decided to venture outside of their laboratory and stage similar studies in the real world. Sleep science hit an all-time high on the strange-ometer when the dynamic duo decided to stage an ambitious experiment with the legendary rock band, the Grateful Dead.

Formed in the San Francisco Bay Area in the 1960s, the Grateful Dead lay at the cutting edge of the West Coast acid culture. They quickly developed a loyal following (affectionately known as 'Deadheads') and played to huge trippy crowds. In 1970 the band's lead guitarist, Jerry Garcia (the inspiration for Ben & Jerry's Cherry Garcia ice cream), happened to bump into Krippner at a party. The two of them chatted, discovered that they were both interested in the potential paranormal power of the mind, and hatched a cunning plan to conduct the world's largest dream telepathy experiment.[4]

In February 1971, thousands of spectators crammed into the Capitol Theatre in New York State to hear the Grateful Dead. What the audience didn't know, however, was that they would be taking part in an ambitious parapsychology experiment. In the middle of the concert several pictures were projected behind the band, along with a message asking the Deadheads to concentrate on the images and psychically beam them across the city.

A few hours before the concert, a psychic named Malcolm Bessent had arrived at Krippner's dream laboratory and settled down for the night. There, the laboratory's researchers were tasked with waking Bessent several times throughout the night and asking him to report his dreams. Publishing the results in the *Journal of the American Society of Psychosomatic Dentistry and Medicine* (just before an article on the personality correlates of venereal disease), Krippner described several striking matches between the images and Bessent's dreams. For example, one of the paintings projected during the gig was called 'The Seven Spinal Chakras', and showed a man deep in meditation and with each of his 'chakras' vividly illuminated. Bessent was woken when the Deadheads were looking at the painting, and reported a dream in which he was 'very interested in . . . using natural energy . . . thinking about rocket ships . . . an energy box and . . . a spinal column'.

After a decade of research, Ullman and Krippner believed that their data demonstrated the paranormal power of dreams. However, when it comes to overturning the known laws of physics, most mainstream scientists are not swayed by a series of experiments emerging from a single laboratory. Instead, it's vital that these studies are successfully replicated by other groups of researchers working at several different establishments. Over the years, many scientists have explored the supernatural side of sleep, and attempted to reproduce Ullman and Krippner's seemingly remarkable results. In fact, one of the most elaborate attempts took place very recently, and I was fortunate enough to be involved.

Do you have the gift?

Carry out the following two experiments to help discover if you have the ability to dream about the future.[5]

Experiment One

1) Choose any national daily newspaper that contains at least twenty pages (but don't buy a copy of it yet).

2) Place your dream diary on your bedside table.

3) When you go to bed, tell yourself that you want to dream about the images and stories that will be printed on page five of your chosen newspaper the following day.

4) Use the techniques from the dream diary to produce an account of your dreams.

5) In the morning, buy a copy of your chosen newspaper, and look at page five. How do your dreams compare to what is on the page? If you are not especially successful with this informal test then you can sleep soundly at night, safe in the knowledge that your dreams can't predict the future. If, however, you seem to be getting more than your fair share of hits, move on to Experiment Two.

Experiment Two

1) For this you will need the dream diary, a six-sided die, a plastic cup, and a friend.

2) Once again, put the dream diary on your bedside table. Before you go to bed, place the die into the plastic cup, place your hand over the top of the cup, and shake the cup. Next, place the cup face down onto your bedside table, so that you don't know which number is showing on the top of the die.

3) As before, dream away and make a note of your dreams in your diary.

4) In the morning, lift the cup and look at the number showing on the die. Use the table below to convert this number into a page number (don't use the number shown on the dice because the earlier parts of newspapers tend to contain major news stories, which might be predictable from existing news reports).

Number on die	Target page in newspaper
1 or 6	Page 5
2 or 5	Page 7
3 or 4	Page 9

5) Buy a copy of your chosen newspaper and give it to your friend. Have them read the entry in your dream diary from the night before and compare it to the information and images on pages five, seven, and nine of the newspaper. Finally, ask them to choose which of the three pages shows the greatest amount of correspondence. Don't influence their choice.

6) If your friend chooses the page that corresponds with your die then count the session as a 'hit', otherwise count it as a 'miss'.

7) Repeat the experiment ten times. If you score six or more hits, get in touch with your nearest parapsychology laboratory.

False awakenings and dream detectives

I can distinctly remember waking up, and the technician asking me whether it was OK to remove the electrodes that were attached to my head. I said that that would be fine, and he started to go to work. A few moments later all of the electrodes were gone, and I was sitting up in bed enjoying my first coffee of the day. It was just then that something strange happened. I heard the door open, and the experimenter walk into the room. I opened my eyes and realized that I had just experienced a false awakening. In reality, I was not sitting up with a cup of coffee, but rather still lying in bed feeling groggy and somewhat confused. Moments later the technician was at the side of my bed for real and started to remove the many electrodes that were glued to my head. I then climbed out of bed, got dressed, and started to drink the cup of coffee that I had dreamt about a few moments before.

I had just acted as a guinea pig in an experiment being carried out by Caroline Watt, from the University of Edinburgh, to discover if it's possible to reproduce some of Ullman and Krippner's classic findings. In addition to investigating the possible existence of telepathy, a significant part of Ullman and Krippner's work had examined whether people's dreams contained information about future events. Watt wanted to discover whether she could replicate this aspect of their research.

The previous evening I had arrived at a sleep centre, been wired up to an EEG machine, and climbed into the laboratory bed. The experimenters had then monitored my EEG data throughout the night. Whenever the data indicated that I had just finished dreaming, I was woken up and asked to describe what had been going through my mind. During the night I had dreamt about being in a cave, a giant inflatable head of Frankenstein, a funeral, and then had the strange false awakening.

Once the EEG sensors had been removed, one of the experimenters explained that she was about to show me a randomly selected film clip. According to the 'dreams can predict the future' hypothesis, my night-time experience should bear an uncanny resemblance to the film. Intrigued, I leant forward and waited for her to press the start button. The target was a film clip of a group of explorers walking through an underground ice cave, and so there did seem to be some similarity between some aspects of my dream and the clip. Perhaps I did have supernatural abilities after all. Or maybe I just got lucky. Either way, it was an interesting experience.

Watt had kindly arranged the session to allow me to discover what it's like to participate in a parapsychological dream experiment, although my result wasn't part of her main study because she was working with volunteers who believed that they might actually have the gift of prophesy. It isn't possible to know whether any one person has paranormal powers from a single session, as any correspondence between their dreams and the film clip might just be due to chance. Instead, scientists test lots of people and then look at the resulting data en masse to see whether it suggests evidence of the extraordinary.

So, how did it Watt's experiment go? Unfortunately, after monitoring about twenty volunteers for several nights on end, the study didn't discover any evidence in support of the supernatural. Watt is not the only one to investigate the paranormal potential of dreaming and draw a blank.

A few years ago I tested top 'dream detective' Chris Robinson.[6] Robinson believes that his dreams can be used to help solve crimes and avert national disasters, and claims to have predicted several terrorist incidents, including the attack on the Twin Towers and the London Underground bombings. The subject of numerous television programmes, Robinson frequently informs the police about his psychic tip-offs.

I investigated Robinson's alleged abilities as part of a television programme looking into psychic detection. We approached Scotland Yard's 'Black Museum', and they kindly loaned us three objects; a scarf used by a rogue milkman to strangle a customer, a bullet that had killed a policeman as he tried to arrest two car thieves, and a shoe that had belonged to a murdered woman. We placed each object in a separate box, and labelled the boxes 'A', 'B', and 'C'. Even though Robinson was unaware of the nature of the objects or crimes, he claimed that he would be able to obtain information about them in his dreams. In addition, two other alleged psychics handled the objects and tried to divine the nature of the crimes.

As a control, a group of my students were asked to look at the objects and guess what had happened during each of the crimes (I have since seen the students' exam results, and can confirm they have no special abilities). A panel of researchers judged the accuracy of their comments, with the results revealing that the psychics performed no better than the students.

And that's the Achilles' heel of Ullman and Krippner's work into paranormal dreaming. Over the years, many researchers have failed to replicate their remarkable findings and, as a result, the work is seen as curious but not proof of the paranormal. This, of course, begs one final question. How is it that so many people claim to have gained a glimpse of the future in their dreams, but yet scientists have struggled to demonstrate this alleged phenomenon in their laboratories? The answer, I suspect, lays in a curious idea known as the 'law of large numbers'.

Every Saturday night something truly amazing takes place in Britain. The chance of this event happening is an astounding fifteen million to one, but nevertheless it takes place as regular as clockwork. People win the National Lottery. Delighted as we all are that a couple of complete strangers have become multi-millionaires on the basis of a tiny outlay, most of us do not see their windfall as evidence of the supernatural. Why? Because we are painfully aware that millions of people buy lottery tickets each week, and that almost all of them lose their money. The National Lottery shows that if enough people attempt to do something improbable, a handful of them are likely to succeed by chance alone. That is the law of large numbers. Sceptics argue that the same explanation can account for anecdotal reports of paranormal dreams. Given that millions of people across the world have several dreams each night, it isn't surprising that some of these dreams will match future events just by chance. Cherry-pick these 'winning tickets' from the mass of unremarkable dreams that never make the headlines, and it looks like something supernatural is taking place.

Dream interpretation: A beginner's guide

While some sleep scientists were investigating the notion of paranormal prophesy, others turned their attention to perhaps the most popular idea about dreaming.

If you go to a party and are foolish enough to mention that you are a psychologist, you almost always hear one of two replies. First, some people say, 'Oh, how interesting. Have you been analysing me?' It's a scary moment for any psychologist, in part because you are probably only seconds away from them telling you all about their fascinating life. The second common response to the 'I am a psychologist' comment concerns dreaming. Once people know that you are interested in the human mind, they frequently describe their latest dream in as much detail as possible, and then ask you what you think it all means. Most psychologists deal with this situation by putting on what I refer to as their 'professional face'. This involves appearing to look interested, while, in reality, desperately trying to remember the telephone number of the cab company that dropped you off at the party.

The notion that dreams contain some form of hidden meaning is extremely popular with the general public, with one recent survey suggesting that around half of all Americans buy into the idea.[7] However, although most people have a vague notion that dream interpretation is somehow associated with Sigmund Freud, they usually do not know very much about the topic. Where, for example, did the idea first come from? Why is the interpretation of dreams so central to Freud's view of the mind? And, perhaps most important of all, is there anything to it? It's time to explore these questions, and much, much more.

The idea of dream interpretation has been around for a very long time. During the second century AD, for instance, the professional diviner Artemidorus Daldianus produced a hefty five-volume work on the topic entitled *Oneirocritica* ('The Interpretation of Dreams'). The first three volumes were intended for public consumption and covered different types of dream content (volume one: 'Anatomy and activity of the human body', volume two: 'Objects and events in the natural world', volume three: 'Miscellaneous'). The remaining two manuscripts were written by

Artemidorus for his son, and described a series of top tips that were designed to help Artemidorus Jr become king of the interpreters. Artemidorus Sr was especially proud of these latter two manuscripts, and cautioned his son against making copies of them in case they fell into the hands of his competitors.

For the most part, Artemidorus adopted a 'one size fits all' approach to dream interpretation, with page after page of *Oneirocritica* carefully cataloguing the meaning of almost every possible symbol and situation. However, Artemidorus also believed that there needed to be some room for manoeuvre, and repeatedly urged wannabe diviners to take into account a dreamer's age, sex, health, social status, height, relationships, occupation, habits, hobbies, and childhood. This multitude of factors could often lead to considerable confusion. In one section of *Oneirocritica*, Artemidorus turns his attention to dreams in which a man has intercourse with his own mother, and notes:

> The case of one's mother is both complex and manifold and admits of many different interpretations . . . The fact is that the mere act of intercourse by itself is not enough to show what is portended. Rather, the manner of the embraces and the various positions of the bodies indicate different outcomes.

Artemidorus then provides a long and detailed account of the interpretations that should be attached to each possible sexual position. For example, instances of face-to-face intercourse. If the dreamer's mother is still alive and his father is in good health, then Artemidorus argues that the dreamer and his father detest one another. If, on the other hand, the dreamer is a farmer, then the dream symbolizes the planting of seeds in inappropriate places, and suggests that the farmer is worried that his crops will fail.

Finally, if the dreamer is a politician, then the dream suggests that he is about to actually do to his country what he has just imagined doing to his mother (that is, make the country happy and take good care of it).

Artemidorus' magnum opus proved to be the surprise hit of its day, and for the following few centuries was viewed as the standard work on dream interpretation. Generation after generation of dream interpreters pawed over the endless copies of the work, and even today Artemidorus' ancient text is still widely available. No doubt encouraged by the enduring success of the *Oneirocritica*, several other dream interpreters have created their own rival systems over the years. Indeed, walk into any New Age shop worth its wind chimes and you will see several books claiming to help you decode dreams and reveal parts of your inner psyche that other types of supernatural mumbo jumbo can't reach. Only one modern-day system of dream interpretation has managed to enjoy the runaway success enjoyed by Artemidorus. This system was created in the 1890s by one of the world's most influential thinkers, and has helped shape thousands of psycho-therapy sessions, and dinner party conversations, ever since. It's time to meet that modern-day master of dream interpretation, Sigmund Freud.

Are you a natural dream interpreter?

Ever wondered whether you would have made a good ancient Greek dream interpreter? Nor me. However, if you are curious, here is a short test based on three scenarios described in the *Oneirocritica*. See how you do.

1) A married woman dreams about sharing her bed with a crocodile and a cat. What would you advise her to do?

2) A man dreams about his canine teeth falling out. What question would you ask him?

3) A perfume dealer dreams that he has lost his nose. Provide three possible interpretations and justify your answers.

Answers

1) According to Artemidorus, 'The crocodile signifies a pirate, murderer, or a man who is no less wicked. The way in which the crocodile treats the dreamer determines the way in which he will be treated by the person who is represented by the crocodile.' In other news, 'The cat signifies an adulterer. For it is a bird-thief.' Therefore the dream suggests that the woman is convinced that her husband is a bad man, and you should advise her to leave him immediately.

2) The *Oneirocritica* states that dreaming about the loss of a tooth often reflects a concern about some sort of house-related payment. However, the teeth can also signify possessions, with molars representing objects of great value, canines denoting worthless rubbish, and incisors indicating household possessions. The loss of the man's canines therefore suggests that he is concerned about debt, or wasting money on a recent purchase. As a result, you need to probe further, asking the man whether he has recently received a large gas bill or bought a cheap sofa.

3) Artemidorus outlined three ways of interpreting this dream. First, the dream may signify that the man is worried about his career. After all, if a perfume dealer cannot smell

then he will lose his business and won't be able to make a living. Second, he may feel guilty about people discovering that he has carried out some type of misdemeanour, with the dream indicating that he is concerned about 'losing face'. Third, the man might be about to die because a skull has no nose. Of course, the dream may reflect a completely different concern, or be entirely random. Give yourself full marks whatever you wrote.

Paging Dr Freud

Born in Austria in 1856, Sigmund Freud developed a fascination with the human mind at an early age. After completing his medical training at the University of Vienna, Freud began to explore ways of helping those suffering from neuroses and other psychological problems. In the mid-1880s he experimented with hypnosis, but struggled to place his clients into trances, and even-tually abandoned the idea in favour of a radically new approach to psychotherapy.

Freud argued that most people have lots of highly sexual and aggressive desires buried deep in their unconscious mind, and that they would feel much better if they managed to get these ideas out in the open. To help with this process, he developed a form of therapy now referred to as 'psychoanalysis'. During a psycho-analytic session a therapist attempts to figure out what is bubbling away in a client's unconscious mind. Some of these insights might be based on the client's everyday behaviour. For example, despite being a life-long smoker himself, Freud argued that smoking is strongly indicative of a constant desire to masturbate. Other times the therapist might ask a client to 'free associate' by lying on a couch and describing the thoughts and images that flow through their mind. Freud was constantly trying to develop new techniques that could be used in psychoanalysis, and wondered whether people's dreams might provide a powerful insight into the unconscious mind.

In the 1890s, the great psychiatrist moved into a house near Grinzing in Austria, and began working on the topic. Freud initially struggled and, in a despairing letter to a fellow scientist, he noted: 'Do you suppose that some day a marble tablet will be placed on the house, inscribed with these words: "In this house on July 24th, 1895, the secret of dreams was revealed to Dr. Sigmund Freud"? At the moment I see little prospect of it.'

However, Freud worked away night after night, and in 1900 he published what was to become one of his best-known works, *The Interpretation of Dreams*. His house in Grinzing was demolished in the 1960s and the empty site is now home to a plaque commissioned by the Austrian Sigmund Freud Society. The plaque reads: 'In this house on July 24th, 1895, the secret of dreams was revealed to Dr. Sigmund Freud.'

According to Freud's classic work, your unconscious mind is not unlike a moody nineteen-year-old who wants to spend their life eating, drinking, and having sex. During the day your conscious mind does a good job of keeping a lid on this bubbling mass of infantile desires, and so all is well. However, at night your conscious mind switches off for a well-earned rest, and so your unconscious impulses start to run amok. This chaos results in a stream of unacceptable thoughts that, if left unchecked, would repeatedly emerge into your conscious mind and wake you up. However, help is at hand. To prevent your conscious mind having to face these unacceptable thoughts, your brain cleverly converts them into less troublesome images and ideas. And, according to the Freudian perspective, it is these more benign images and ideas that form the basis of your dreams. Seen in this way, dreams are essentially 'guardians of sleep' that have evolved to prevent your innermost infantile desires emerging from your unconscious and keeping you up all night.

During psychoanalysis, a Freudian therapist attempts to uncover their client's desires by having the client describe what they actually dreamt about (referred to as the 'manifest content' of a dream) and then using this to determine the unfulfilled need that the dream represents (referred to as the 'latent content'). Although Freud thought that dreams could reflect many different types of hidden needs, much of his work focused on two very basic drives: sex and aggression.

Freud discussed several techniques for analysing dreams, but two of them have stood the test of time.

First, Freud was a fan of asking the client to 'free associate' by saying what thoughts and images run through their head when they think about their dream, and then using these comments to uncover their hidden desires. Second, Freud also followed in the footsteps of Artemidorus and believed that some of the characters, objects, animals, and scenarios that frequently appeared in dreams have set meanings. The majority of these alleged meanings are often somewhat counter-intuitive and frequently sexual. If, for example, a man dreams about smoking a cigar then, according to some Freudians, he is worried about the size of his penis. If he dreams about having sex with his mother, then he is in love with his father. And if he dreams about having sex with his mother while smoking a cigar, then you can bet your bottom dollar someone is in for an embarrassing night. Just kidding. In fact, some Freudians argue that circular objects often represent vaginas, that almost any process that involves separating a part from the whole (such as losing a tooth) is indicative of a fear of castration, and that riding a merry-go-round is a sure sign of a strong need for sex. Although this symbolic approach to dreaming is popular with many psychoanalysts, Freud himself was not overly keen on this approach.

Two examples from Freud's writings illustrate his somewhat subjective, and often sexual, approach to dream interpretation.

In one instance, a female German client reported a dream in which her husband suggested that they tune their piano, and she replied that it wasn't worth it because the hammers needed reconditioning. When the woman described the dream to Freud, she said that the piano was a disgusting old box that made a horrible noise. Freud seized on these seemingly chance remarks, noting that in German the word for box ('*kasten*') is similar to the word for chest ('*brustkasten*'), and concluded that the dream concerned the woman's deep-seated worries about her physical development during puberty. Further analysis supposedly revealed that the woman began to be disappointed with her body around this time, and had an unconscious desire for larger breasts.

During another session, a male client described a dream that involved him seeing a house between two stately palaces. The doors of the house were closed. The man's wife then led him along the street, pushed open the doors to the house, and encouraged him to go inside. To Freud, this was an easy bowl. According to the great psychoanalyst, dreams involving people walking up narrow spaces and opening closed doors were almost always sexual in nature. As a result, Freud was convinced that the man's dream reflected his unconscious desire to have sex with his wife, and that the two stately palaces represented her buttocks.

Are you a natural Freudian?

Freudian symbolism often seems fixated on sex, and includes more than 700 symbols that allegedly represent the penis, more than 100 for the vagina, about 500 for intercourse, and 25 for masturbation. Critics believe that these lists are the result of Freudians having a dirty mind. However, on the upside, the lists allow us to construct a fun test that will reveal whether you have an innate ability to see the world through the eyes of a psychoanalyst.

Take a look at the ten items in the following table (without peeking at the answers opposite). According to many Freudians, when these objects and ideas crop up in a dream, they represent a penis, vagina, intercourse, or masturbation. Please place a cross in one of the columns in each row to indicate how you think each item should be classified.

	Penis	Vagina	Intercourse	Masturbation
1) Riding a horse				
2) A pipe				
3) A pocket				
4) A balloon				
5) A church				
6) Playing an instrument				
7) Going down a ladder				
8) Running inside a house				
9) A castle				
10) A suitcase				

Here are the answers. Award yourself one point for each item that you classified correctly.

	Penis	Vagina	Intercourse	Masturbation
1) Riding a horse			x	
2) A pipe	x			
3) A pocket		x		
4) A balloon	x			
5) A church		x		
6) Playing an instrument				x
7) Going down a ladder			x	
8) Running inside a house			x	
9) A castle		x		
10) A suitcase		x		

If you scored more than eight then you clearly have an innate understanding of the Freudian approach to the human mind. Scores of between three and seven suggest a more middling ability. If you scored less than three then you are clearly repressing your thoughts about genitalia, masturbation, and intercourse.

The Freudian approach to dream interpretation quickly attracted a considerable amount of scepticism. After meeting Freud in 1909, the great American philosopher William James wrote: 'I confess that he made on me the impression of a man obsessed with fixed ideas. I can make nothing with his dream theories, and obviously "symbolism" is a most dangerous method.'

Over the next forty years or so, the debate between psycho-analysts and their critics grew increasingly bitter, with sceptics arguing that Freud's approach to dream interpretation was inher-ently subjective, difficult to test, and often wrong.

One study, conducted by Michael Schredl from the Central Institute of Mental Health in Germany, involved asking people who were into dream interpretation to rate how beneficial they had found dream symbolism versus non-Freudian approaches.[8] The results revealed that the non-Freudian approaches were significantly more beneficial. Other research demonstrated how Freudian dream analysis can manipulate people's memories of their childhood.[9] In addition to interpreting people's dreams, many psychoanalysts claim that they can use them to uncover repressed memories, with many of these alleged memories relat-ing to traumatic events that took place during people's childhood. But are these memories genuine or fictitious? In a study conduc-ted by Giuliana Mazzoni from the University of Florence, more than one hundred volunteers first completed a questionnaire asking whether they had ever been lost in a public space or aban-doned by their parents. Two weeks later the volunteers that had indicated that they had not had either of these experiences were asked to attend a 'dream interpretation' session with a senior clin-ical psychologist. During this session, the clinician explained that dreams contain hidden Freudian symbols, and asked the volunteer to describe a recent dream. Whatever the content of volunteer's dream, the clinician explained that their report suggested that they had repressed memories of a difficult childhood experience, such as them getting lost in a public place or being abandoned by their parents. About two weeks later the volunteers were interviewed about their childhood a second time. Remarkably, the majority of them explained that they were certain that they had indeed experienced the events suggested by the clinician. Mazzoni's study suggests that Freudian dream interpretation doesn't uncover repressed memories, but rather results in reports of events that never actually happened.

After reviewing these sorts of studies, sleep scientist William Domhoff from the University of California concluded: '. . . there is no reason to believe any of Freud's specific claims about dreams and their purposes'.[10]

But do the great psychoanalyst's ideas really amount to little more than a pipe dream? Are dreams, as some have argued, the 'meaningless froth on the beer of sleep'? After spending night upon night studying thousands of volunteers across the world, sleep scientists discovered that Freud was right in thinking that dreams contained hidden messages, but wrong to believe that they were almost always to be reduced to sex and violence. Over time these researchers began to piece together a radically new theory of dreaming. This approach is supported by the results of countless scientific studies and, perhaps most important of all, has helped thousands of people to improve their lives. And that's where we are heading next.

ASSIGNMENT

Faking Freud

Although there is very little evidence to support the idea that certain types of dreams have fixed meanings, people are fascinated with the topic and frequently mention it at parties. Rather than dismiss the issue out of hand, it's often fun to appear to be able to offer a deep insight into people's dreams (this is especially true if you happen to be single, and find the dreamer somewhat attractive). Here are some common interpretations for many of the most frequently experienced dreams. Whenever someone says that they have had one of these dreams, feel free to use the information below to 'interpret' it. After you have offered your deep and meaningful insights, do remember to explain that there is no scientific evidence to support anything that you have said.

'I was being chased'
Dreams that involve any form of chase indicate a sense of fear or threat. Ask the dreamer whether there's a family member, friend, or work colleague that they find especially threatening. Then nod knowingly.

'I felt like I was falling'
This category consists of dreams that involve any sense of downwards motion, including dreams in which the dreamer falls to earth from outer space, tumbles down a mountain, or drops down any kind of shaft. These dreams indicate anxiety about a loss of status, security, money, job, or relationship.

'I was flying through the air'

Flying dreams can take many different forms, including soaring through the air like a bird or simply levitating above the ground. No matter what form they take, these dreams indicate that something very positive is happening in the dreamer's life, that they feel in control of an important situation, or that they are looking forward to a future event.

'I was completely naked in public'

The 'naked in public' scenario is surprisingly common, and suggests that the dreamer is worried that their friends, family, or work colleagues will discover a hidden truth. Either that, or they are just concerned about being naked in public.

'I was struggling with some sort of test, interview, or performance'

This type of dream can take many different forms. The dreamer might, for example, struggle to complete an exam because their pen doesn't work, or the questions don't make any sense. Alternatively, they might have to give a talk, be interviewed, or appear in a play, but face a hostile audience, difficult questions or can't remember their lines. All of these dreams suggest that the dreamer doesn't feel prepared for an important challenge in their lives.

'My teeth were falling out'

The loss of teeth in a dream is surprisingly common, and often reported by dreamers who suffer from low self-esteem. It may also reflect concerns about aging. Or it may mean something completely different. Who knows?

Lesson 7

DREAM THERAPY

Where we uncover the secret life of dreams, meet the world's greatest therapist, and discover how you can find the solution to any problem in your sleep.

In this lesson we are going to continue our journey into dreamland by finding out what really goes on in your dreams, and how you can use these strange nocturnal experiences to improve your life and change the world. However, before we begin, please take a few moments to try to solve the following three puzzles. If you can't think of a solution, don't worry.

1) The following letters form the start of an infinite sequence. What are the next two letters in the sequence?

O, T, T, F, F, . . .

2) What is the rule that explains this sequence of numbers?

8 5 4 9 1 7 6 3 2

3) Can you work out the next letter in this sequence?

CYWOTNLIT

Many thanks. More about these puzzles later.

The secret life of dreams

Within just a few years of sleep scientists discovering the remark-able relationship between dreaming and REM, researchers across the globe had wired up thousands of volunteers to EEG machines, repeatedly woken them up throughout the night, and asked them to describe the thoughts and images that were going through their head. These early studies quickly yielded several startling insights into the hitherto hidden world of dreaming, including the fact that almost everyone has about five dreams each night (but often think they dream far less because they tend to forget their dreams in the morning), that dreams tend to become longer as the night wears on (the first dream of the night often only lasts just a few minutes, while the final dream can hit the forty-minute mark), and that dreams happen in 'real time' (if people are woken after spending five minutes in REM they will describe about five minutes of action). Thrilled by this rapid progress, researchers then started to focus their attention on the content of people's dreams. Leading the charge was sleep scientist William Domhoff.

Domhoff is a fascinating man. Born in 1936 and spending most of his career working at the University of California, he has made a name for himself in two very different areas of aca-demia. In the late 1960s, he published *Who Rules America?*, a highly controversial and bestselling book arguing that America was con-trolled by corporations and banks. However, when Domhoff was not studying how the rich run America, he focused his attention on the stuff of dreams. Over the years he has collected, and analysed, thousands of dream reports.[1] Domhoff treats each of these reports like a little play that involves a cast of characters (friendly man, aggressive woman, sexy stranger), a setting (inside a house, outside in a field, down a mineshaft), and a plot (a man comes home and watches television, a television comes home and watches a man).

Domhoff's work revealed several surprises about the secret life of dreams. Because you only tend to remember your more bizarre dreams, you might be tempted to think that all of your nocturnal fantasies are weird. In fact, about 80 per cent of your dreams tend to involve fairly ordinary scenarios, such as being at work, doing the washing-up, or walking the dog. Domhoff also discovered that your nocturnal adventures tend to be populated by people you know, with your family members cropping up in about 20 per cent of your dreams, and friends being present about half of the time. On the very few occasions that a celebrity pops up in your dreams, they usually make a fleeting appearance rather than playing a major part. Domhoff also discovered that you cast yourself in the starring role in almost all of your dreams, and that the vast majority of nocturnal adventures are experienced from a first-person perspective.

Domhoff's work has provided strong support for an idea that sleep scientists refer to as the 'continuity hypothesis'. According to this theory, most dreams are a continuation of what is happening in your everyday life. If, for instance, you spend most of your day in the office, then you will tend to have lots of desk-based dreams. However, head off to the beach for a couple of weeks, and suddenly your dreams become full of sand, sea, and sun. There is, however, one important difference between your dreams and everyday experiences. Regardless of what's happening in your waking life, around 80 per cent of your dreams involve some type of negative emotion. In fact, the murder rate in your dreams is higher than in any city in the world, and the most common dreams nearly all involve a sense of fear, stress, or anxiety. Whether you are struggling in an exam or being chased along a street, falling down a toilet or missing a train, being naked in public or fighting off a zombie apocalypse, dreamland is a hotbed of negativity.

To make matters worse, this high level of negativity goes through the roof if you become stressed in your waking life. When, for example, women are struggling to deal with a difficult divorce they often have highly anxious dreams about loneliness and fear. Similarly, when Vietnam veterans returned home, they frequently experienced dreams associated with guilt and violence.[2] Even hearing about a stressful event can trigger highly traumatic dreams, with one study showing that the attack on the Twin Towers caused a dramatic increase in the number of people across America dreaming about explosions, death, and fire.[3]

Intrigued, many sleep scientists began to wonder whether this overwhelming incidence of far-from-sweet dreams has evolved because they are, in some strange way, good for you.

Dirty dreams, wet dreams, and animals

Several sleep scientists have built on Domhoff's ground-breaking work by examining specific types of dreams in considerable detail.

For instance, Michael Schredl from Germany's Central Institute of Mental Health, analysed reports from one twenty-two-year-old man over several years, focusing on dreams in which the man had gone to the toilet.[4] Schredl discovered that dreamland is not unlike the centre of several cities on a Saturday night, with the young man frequently having to urinate in a can, being forced to use filthy toilets, and often being disturbed mid-defecation by shouts of abuse.

Other researchers have focused on sex, with their work revealing that about a third of dreams involve some sort of sexual content, but that actual intercourse only takes place about 10 per cent of the time. Calvin Kai-Ching Yu from Hong Kong Shue Yan University has carried out a detailed study of sexual dreams in men.[5] In his classic paper 'Sex Dreams, Wet Dreams, and Nocturnal Emissions', Yu reports that 5.2 per cent of men have kissed a monster in their dreams, 3.4 per cent have indulged in foreplay with an animal, and 1.7 per cent have had intercourse with an 'object, plant or rock'. In other news, 10.3 per cent of men have dreamt about having sex with their mother and 6.9 per cent have had an overly good time with their elder sister.

Finally, sleep scientist Robert Van de Castle has focused on our four-legged friends.[6] Van de Castle collected hundreds of run-of-the-mill dream reports and split them into four categories: those containing no animals, those containing the same number of animals and humans, those with more humans than animals, and those with more animals than humans. Van de Castle then looked at the aggressive content of the dreams. When there were no animals about, 45 per cent of the dreams involved some form of aggression. However, drop in an animal and this rises to a scary 60 per cent. If there are more animals than humans, there is a massive 80 per cent chance of some sort of aggressive behaviour kicking off. Van de Castle also discovered that animals tend to attack humans more than twenty times more often than they are friendly towards them. In short, if an animal wanders into your dreams, run.

The twenty-four-hour shrink

Why should your dreaming mind be so fixated on scary thoughts and images? After much head scratching, several sleep scientists speculated that these negative scenarios weren't designed to terrify you, but rather to help you cope with your everyday concerns and worries. Various versions of this 'dreams as nocturnal therapist' theory were proposed. Some researchers noted that negative events tend to lose their emotional impact when they are repeatedly experienced, and speculated that dreaming about these difficult life events might lessen the trauma. Others suggested that dreams might help you cope with a present-day problem by replaying memories of past events that evoked similar emotions. Whatever the exact explanation, it was an intriguing theory and researchers soon set about putting the idea to the test.

In one study, volunteers were first shown an unpleasant film containing scenes from a bloody autopsy.[7] They were then randomly allocated to one of two groups, and spent the night in a sleep laboratory. The volunteers in one of the groups were woken up whenever they started to dream, while those in the other group were woken up the same number of times when they weren't dreaming. As a result, all of the volunteers obtained the same amount of sleep, but those in one of the groups were deprived of dreaming. The following morning all of the volunteers were shown the autopsy film a second time, and asked to rate how anxious they felt. According to the 'dreams as therapist' theory, the volunteers that were allowed to dream the night away would have had an opportunity to deal with the anxiety caused by the first showing of the film, and so not feel so stressed second time around. And, that is exactly what the researchers found.

Encouraged by these types of findings, sleep scientist Rosalind Cartwright, from Rush University Medical Center, decided to put the 'dreams as therapist' theory to a very different type of test.[8] When people attend several psychotherapy sessions they tend to feel better as time goes on. Cartwright reasoned that if the same idea applied to people's nightly sessions with their 'nocturnal therapist', dreams that take place towards the end of the night should feel significantly more upbeat than those that occur earlier in the night. To find out if this was the case, Cartwright invited a group of volunteers to her sleep laboratory, woke them up from REM at different points throughout the night, and asked them to rate how anxious their dreams felt. As predicted, the later dreams were far more emotionally positive than the earlier ones.

So far, so good, but would the theory stand up to scrutiny in a more realistic context? Rosalind Cartwright decided to explore the idea with the help of a group of female divorcees who had been diagnosed with depression.[9] At the start of the experiment, Cartwright had her volunteers spend a few nights in her sleep laboratory. The researchers repeatedly woke up the women when they had just finished dreaming, and asked them to describe what had been going through their mind. Each of these reports was then carefully analysed, with researchers rating how emotional the dream seemed and whether it featured the volunteer's ex-spouse. A year later, the researchers contacted the volunteers to find out if they were still depressed. Finally, the research team looked at the volunteers' dreams from the start of the study to discover any differences between the women that had, and hadn't, recovered from depression. The results were remarkable. As predicted by the 'dreams as therapist' theory, the volunteers who had started the study having more emotionally laden dreams, and dreams that featured their ex-spouse, were significantly more likely to recover from depression. Unfortunately this doesn't mean that maxing out on dreaming will help alleviate depression because, as is often the case in life, other work has shown that it's possible to have too much of a good thing (see below).

Time and again, the results of these types of studies showed that dreaming plays a vitally important role in coping with the stresses and strains of everyday life, and so supported the 'dreams as therapist' theory. Armed with this new and exciting theory of dreaming, several sleep scientists started to explore how the idea could be used to improve people's waking lives.

Dreaming and depression

In Lesson 1 we discovered that deep sleep is almost the exact opposite of REM. During REM you are surprisingly close to wakefulness and your brain is highly active, whereas in deep sleep you are cocooned away from reality and your brain is far more dormant. The two states benefit you in different ways, with REM helping you to cope with your worries and concerns, and deep sleep restoring energy levels to your brain and body. Because of this, some researchers believe that there is a trade-off between the amount of time that you spend in each state during the night. Spend lots of time in deep sleep and you will face the day feeling physically great, but will be more likely to struggle with whatever is on your mind. Alternatively, spend most of the night dreaming and you will be better able to deal with the slings and arrows of outrageous fortune, but will feel less energetic.

People who have been diagnosed with severe depression often struggle to get a full night of sleep and, even when they do, frequently feel as if they have very little energy in the morning. Scientists wondered whether this might be due to their brains trying to deal with the stress of their waking life by spending too much time in REM and, as a result, obtaining a relatively small amount of deep sleep. To find out if this is the case, researchers invited volunteers who were suffering from depression into sleep laboratories and spent the night monitoring their brain activity. The results provided remarkable support for the theory.[10] Most people have their first dream around ninety minutes after they have gone to sleep, and this initial dream tends to last around about ten minutes. In contrast, those suffering from depression tend to start to dream just forty minutes after they have fallen asleep, and this dream can last as long as twenty minutes. This trend continues throughout the night, with the dreams of depressives tending to be longer, and more frequent, than those not suffering from depression. As a result, they spend very little time in deep sleep and so wake up feeling totally exhausted.

According to the 'too much time in REM robs you of energy' theory, it should be possible to help those diagnosed with severe depression by having them cut down on the excessive amounts of time that they are spending with their 'nocturnal therapist'.[11] Research carried out by Gerald Vogel, from Emory University, suggests that this might well be the case.[12] Vogel invited depressed patients from a psychiatric hospital to spend several nights in his sleep laboratory. The research team randomly allocated the patients to one of two groups, and monitored their brainwaves throughout the night. The patients in one of the groups were woken up whenever they started to dream, while those in the other group were woken up the same number of times when they weren't dreaming. As a result, all of the patients obtained the same amount of sleep, but those in one of the groups were deprived of dreaming. It wasn't easy. Preventing people from dreaming quickly produced a rebound effect that caused the patient to start to dream the moment they fell back asleep. However, Vogel persevered and eventually developed a schedule that involved the patients being woken up for a maximum of thirty times a night for six nights, and then sleeping normally during the seventh night (I refer to this as 'the day of rest'). After just three weeks, half of the dream-deprived patients were judged well enough to leave the psychiatric hospital. In contrast, none of those that had been woken up when they weren't dreaming showed any significant improvement.

Unfortunately, Vogel's innovative treatment is extremely difficult, if not impossible, to maintain for extended periods of time. However, many antidepressants reduce the amount of time that people spend dreaming each night (as well as increasing the vividness of those dreams that do occur), causing some researchers to conclude that this may well be one of the main reasons why these medications are effective.[13]

Dream therapy

In the 1960s, sleep researcher William Dement was smoking two packs of cigarettes a day and had developed a chesty cough.[14] One day he coughed into a white handkerchief and saw tiny flecks of red in his sputum. Concerned, he went to his doctor and was immediately rushed into hospital for an X-ray. The following day, Dement was back in his doctor's office looking at a light box. The X-ray being displayed on the box clearly showed that both of his lungs were riddled with cancer, and Dement realized that that he was staring death in the face. Not surprisingly, he was distraught. Despite knowing the health risks associated with smoking, he had ignored the warnings and was now paying the price. It was then that he woke up. In reality, Dement *was* a very heavy smoker, but the blood, X-ray, and results were entirely imaginary. Nevertheless, the dream had a considerable impact on his life. Concerned that it reflected what might come to be, Dement immediately gave up smoking and has never touched a cigarette since.

Intrigued by these types of stories, researchers wondered whether people would benefit from taking a closer look at their dreams in order to understand more about the issues that their imaginary 'nocturnal therapist' was working on. Like Freud, these researchers were convinced that people's dreams did indeed contain some form of hidden meaning. However, unlike the cigar-loving psychoanalyst, they didn't buy into the notion that dreams were usually about people's unconscious sexual urges, that certain dream images had fixed meanings, or that decoding dreams required years of training. Instead they based their work on the findings from sleep science, assuming that dreams reflect people's everyday concerns, that dream images mean different things for different people, and that gaining an insight from your dreams can be both quick and simple.

Perhaps the most widely used way of helping people to get in touch with their 'nocturnal therapist' has been developed by psychotherapist Clara Hill (see overleaf).[15] In the first part of this simple three-stage process, the psychotherapist helps their client describe a recent and memorable dream in as much detail as possible. Second, the client is then encouraged to think about how the dream relates to real events in their life. Finally, the psychotherapist and client work together to explore the implications of these insights for the client's waking life. The effectiveness of Hill's procedure has been validated in several studies. In one experiment, for instance, Hill assembled a group of people who were suffering from various psychological problems, and randomly allocated each of them to one of two groups. The volunteers in both groups then received the same type of psychotherapy, but those in one of the groups also carried out Hill's dream work.[16] Compared to those just receiving the psychotherapy, the volunteers who undertook the dream work rated their sessions as significantly more interesting, useful, and insightful. Additional studies have shown that dream work can help people deal with a range of psychological problems, including low self-esteem,[17] relationship issues,[18] and depression.[19]

Dream work

Making an appointment with your 'nocturnal therapist' isn't difficult. In fact, all you need to do is carry out the following three-stage procedure.[*]

---

[*] The procedures described in this section are based on work carried out by psychotherapist Clara Hill. These exercises are designed to provide a general insight into the sorts of techniques that are used by health professionals. If you believe that you, or your child, have a psychological problem, please consult a professional.

1) Exploration

What happened?

Note down your dream in as much detail as possible. Write your description from a first-person perspective ('I can re-member being in a lift . . .'), and feel free to add some drawing and diagrams. Don't worry about describing how the dream made you feel, because this stage is all about recording exactly what happened. You might find it helpful to think about your dream like a film, and make a note about the characters involved, the setting, and the plot. After you have produced your description, imagine that your dream is actu-ally going to be made into a film – what would the film be called?

Example

I can remember being in a lift that was going up. Suddenly the lift stopped and the doors opened. There was a giant dragon outside the lift, and it wanted to get in. There clearly wasn't enough room for the two of us in the lift, and so I tried to close the door. However, every time the doors started to close, the dragon managed to get his foot inside and they opened again. This went on for quite some time.

I would call my film 'Dragon-lift'.

How did your dream make you feel?

Now it's time to turn to the more emotional side of your dream. How did you feel during the dream? You might find it helpful to take a look at the following list of emotions, and pick out the ones that describe how you felt.

discovery, confusion, surprise, awe, wonder, happiness, sadness, amusement, joy, courage, pride, fear, anger, frustration, timidity, cowardice, pity, modesty, shame, detachment, boldness, patience, relaxation, stress, envy, nervousness, security, love, hatred, despair, confusion.

Example

I felt extreme frustration and anger during my dragon dream. The dragon had no right to try to push his way into my lift, and he kept on trying again and again.

Associations

Does your dream seem to relate to something that is happening in your life now, or that you have been thinking about recently? Or perhaps something that happened in your past? Did a recent event trigger your dream?

Example

I'm not sure. The dream took place in an office lift and I have been struggling at work. My boss was very critical of my performance a few days before the dream happened.

2) Insight

Let's assume that your dream has some meaning. What messages might be hidden in the dream? You might find it helpful to think about what the dream might be saying about your personal life, career, personality, or relationships.

Example

I think the dream relates to my career. The dragon represents my boss. The lift is all about me being promoted. I think that maybe my boss is actively trying to stop me being promoted. I suspect that he knows that there will only be a handful of promotions this year, and he wants to get one of them. That is why he was trying to get into the lift, and that is why he was trying to stop me going up.

3) Action

Imagine that you could change your dream in any way you like. What would you change? For example, how else could the dream end? Who else might be involved? Imagine that your new dream is being made into a film – what would it be called? Feel free to be as creative as you like. Finally, what does this new and improved dream say about how you might change your waking life?

Example

I would like to have had the strength to push the dragon out of the lift, and hit the 'up' button. Then, in my ideal dream I would have gone to the very top of the building, and walked out into an amazing office with great views and people working away on a creative project. I would call my new dream 'Defeating the dragon'. I guess the dream is suggesting that I need to deal with my boss, and that I am not happy where I am in the organization. I have always been the ambitious type, and maybe now is the time to move onwards and upwards.

Hill's approach has been shown to be highly effective, but it can take a considerable amount of time to use. Those after a quicker fix might want to consider a related, but easier, technique that has been developed by psychologist and therapist, Teresa DeCicco, from Trent University in Ontario (see overleaf). This procedure, known as 'The Storytelling Method', involves identifying the main elements in your dream, thinking of words that you associate with these elements, forming a story around these associations, and then relating this story to your life. DeCicco's technique has also been subjected to a considerable amount of testing, and has been shown to help people obtain a significant insight into their lives.[20]

Exciting as these scientifically supported methods of dream interpretation are, when it comes to the power of dreams, they are just the tip of the iceberg. Your 'nocturnal therapist' doesn't just help you identify issues in your life, but can also help you to find new and innovative solutions to your problems.

The Storytelling Method

If you want to try to figure out how a dream might relate to your life, learn to use 'The Storytelling Method'.[*] It consists of the following six stages:

1) Write down a description of your dream in as much detail as possible. Try to write in short sentences, keeping the dream in order.

   I was in a lift.
   There was a dragon trying to get in.
   The dragon kept stopping the lift doors closing with his foot.

---

[*] The exercises described here are designed to provide a general insight into the sorts of techniques that are used by health professionals. If you believe that you, or your child, have a psychological problem, please consult a professional doctor or clinical psychologist.

2) Underline the word(s) or short phrase(s) in each sentence that seem most important to you.

I was in a <u>lift</u>.
There was a <u>dragon</u> trying to get in.
The dragon kept <u>stopping the lift doors closing</u> with his foot.

3) Draw a two-column table and write 'List A' at the top of one column and 'List B' at the top of the other. Write each of the words or short phrases that you identified in step 2 under List A.

List A	List B
lift	
dragon	
stopping the lift doors closing	

4) Write the first one or two words that come to mind when you think about each of the words or phrases in List A. Make a note of these associations in List B.

List A	List B
lift	*up*
dragon	*power*
stopping the lift door closing	*prevent*

5) Form a short and simple story around the words in List B. Keep the words in the order that they appear in the list, and ensure that the story makes sense.

I want to go up, but a powerful force is preventing me.

6) Take a look at your short story. Does this story relate to an aspect of your life in some way? If so, does the story

give you an insight into that aspect? If yes, write about the insight and how it relates to your life.

I have been trying to get a promotion at work, and I think that my boss has been blocking it. This is obviously playing on my mind, and it is time that I did something about the situation.

Dream solutions

In the 1970s, William Dement, a sleep researcher at Stanford University, speculated that dreams often represent a weird form of lateral thinking in which the mind attempts to solve a problem by looking at it from several new and unusual perspectives. To find out whether he was right, Dement assembled a group of 500 volunteers and presented them with the type of puzzles that you saw at the start of this Lesson (page 221).[21] These puzzles were carefully chosen because they appear tricky at first, but then the solution becomes obvious when you view the puzzle from the correct perspective.

Half of Dement's volunteers were shown the puzzles in the morning, and were asked to try to solve them that evening. The other half was given the puzzles just as they went to bed, and they were asked for their answers in the morning. According to Dement's theory, the group that went to bed would outperform the daytime puzzle solvers – which is exactly what happened, with those who slept on the problem obtaining significantly higher scores.

Not only that, but many of the dreams related to the puzzles.

For example, the first problem involved finding the missing letters in the sequence 'O, T, T, F, F, . . .'. One participant described how they had dreamt about wandering through an art gallery counting the paintings. All was well, except that the sixth and seventh paintings had been ripped out of their frames. They woke up and realized that the sequence is derived from the initial letter of the words 'One', 'Two', 'Three', and that the missing letters relate to the words 'Six' and 'Seven'.

Oh, and if you haven't solved them yet, the numbers in the second puzzle are in alphabetical order ('E' is for 'Eight', 'F' is for 'Five' and 'Four', and so on), and the letters in the final puzzle are the first nine letters of the words in the sentence 'Can you work out the next letter in this sequence?' So the next letter is 'S'.

In a follow-up study, Gregory White from the Redding Academic Center in California examined whether this effect might be due to relaxation. White asked a group of volunteers to make a list of up to eight 'moderately distressing' personal problems.[22] Some of the volunteers were then asked to explore possible solutions to their problems after either dreaming, or carrying out a relaxation exercise. The volunteers carried out these procedures for ten days, recording both the degree to which they had found a solution and their level of distress. Compared to the volunteers carrying out the relaxation exercises, those who sought solutions in their dreams solved more of their problems, and reported significantly lower levels of stress.

Another study, conducted by Denise Cai from the University of California at San Diego, showed that it is dreaming, rather than sleeping, that helps people solve problems.[23] In this study, a group of volunteers were presented with a series of puzzles that required creative thinking, and then asked to lie down for just over an hour. Some of the volunteers rested while others fell asleep. Among the sleepers, some dreamt and others didn't. At the end of the session everyone was presented with the puzzles a second time, and the results revealed that only those who had spent some time dreaming solved more.

Work by Sara Mednick from University of California, San Diego, showed that even a few minutes' dreaming had a considerable impact on people's problem-solving skills.[24] Mednick first tested volunteers' creative problem skills by presenting them with sets of three seemingly unrelated words ('elephant', 'lapse', 'vivid') and asking them to think of a fourth word that relates to all of them ('memory'). The volunteers were then randomly split into three groups. One group sat quietly in a chair, another group took a sixty-minute nap, and the third group was allowed the luxury of a ninety-minute snooze. Because of the ninety-minute sleep cycle, only those in this third group had a dream. Finally, the volunteers' creative problem-solving skills were then tested a second time. Only those who napped for ninety minutes obtained higher scores in the test.

There is considerable anecdotal evidence suggesting that the same type of dream problem solving can also help people find the solution to important issues in their lives. Take, for example, how a dream persuaded entrepreneur Kally Ellis to completely change her life. In 1990, Ellis was in a job she hated and had just broken up with her long-term boyfriend. One night, Ellis had a dream in which she saw herself running a trendy flower shop, and exhibiting her beautiful bouquets and arrangements across the world. She woke up feeling a wonderful sense of happiness, and knew that her dream contained an important message. Ignoring the recession and her complete lack of experience in floristry, Ellis handed in her notice and opened a small flower shop in Shoreditch, London. After four years of hard work, she got her first big break when she was asked to provide the flowers for a *Vanity Fair* event. All went well, and now her business supplies arrangements for the Oscars, the Cannes Film Festival, and several top London hotels.

Ellis is not the only one to have been influenced by the night. A few years ago I asked people whether they had ever experienced a dream that had changed their life and, if so, to describe what had happened. Hundreds of people sent in their life-changing dreams. One woman described how she had been overweight, but in her dream saw a new her that regularly went to the gym and stayed away from the dessert trolley. Impressed with this imaginary slimline version of herself, the woman decided to adopt a much more healthy lifestyle and lost a large amount of weight. In another instance, a man described how he had had a dream in which he was being a much better father to his children. In the dream, he had decided to work fewer hours in the office and instead spend more time at home. As a result, both the man and his family were much happier. The man woke up, went to work, and negotiated a part-time position.

Both Dement's research, and these types of experiences, suggests that your 'nocturnal therapist' doesn't just help identify your worries and concerns, but can also help you to see new and innovative solutions to your problems. Harvard sleep expert Deidre Barrett decided to take the idea one step further by seeing whether it was possible to persuade people's 'nocturnal therapists' to tackle a specific real-world problem.[25]

Barrett assembled a group of volunteers who were facing a major life decision, such as a career change or relationship issue. She then asked them to jot down the problem before they went to sleep (see page 243). Each morning the volunteers were asked to make a note of whether their dreams helped them solve the issue. About 50 per cent of the volunteers dreamt about their problem, with a notable 70 per cent of those finding a solution in their dream.

In one instance, a young woman named Mary had been offered a place on both clinical psychology and industrial psych-ology courses, and couldn't decide which to choose. In her dream, Mary saw herself flying over a map of America. The pilot announced that the plane had engine trouble and needed to land. Mary suggested landing in Massachusetts but the pilot said that it was very dangerous, and that the plane needed to head further west. Mary woke up and thought about the dream. The clinical psychology course was based in her home state of Massachusetts, whereas the industrial course was in California. Mary realized that deep down she didn't want to stay so close to home, and that moving away was the right decision.

In another example, a young man named Frank was trying to decide whether he should re-join his local softball team. Frank was afraid that the practice sessions would distract him from his studies and, as a compromise, had considered just going to watch the games as a spectator. During the study Frank repeatedly dreamt about standing around some tents that didn't go all the way to the ground, and awkwardly staring at what was going on inside each of the tents. These dreams made Frank feel very uncomfortable and reminded him of the phrase 'a watcher rather than a doer'. This phrase had very negative connotations for Frank and as a result he decided that he wouldn't be happy simply observing the softball games. A week after the study Frank signed up for the team.

This research shows that it is possible to ask your 'nocturnal therapist' to focus their attention on a specific problem. Then, throughout the night they look at the issue from a variety of perspectives and, all being well, you see their fruits of their labour in a dream. This simple procedure can radically improve people's lives.

Finding your dream solution

Do you want to use your dreams to help solve a problem? Try the following technique.*

- Place a pen, a pad of paper, and a torch on a table next to your bed.

- Briefly jot down your problem on the paper before you go to bed. If possible, try to find some objects that are connected to the problem and place them on the table. For example, if you are trying to tackle a problem at work you might want to place a folder containing relevant documents on the table. If you are trying to find your way through a difficult relationship, consider placing your wedding ring there.

- When you are tucked up in bed, imagine yourself dreaming about the problem, waking up, and writing a solution on your notepad.

---

* The exercises described here are designed to provide a general insight into the sorts of techniques that are used by health professionals. If you believe that you, or your child, have a psychological problem, please consult a professional doctor or clinical psychologist.

- Just as you are falling asleep, mentally tell yourself that you want to dream about the problem and find some possible solutions.

- If you wake from a dream during the night, make a quick note of the main gist of the dream on your pad. Invite more dreams and go back to sleep.

- When you wake up in the morning, lie quietly for a few minutes before you get out of bed. Can you remember any of your dreams? If so, note down the main themes and images that went through your head.

- After a week or so, review your dreams. What are the main themes and images to emerge? Did any dream seem especially striking and helpful?

The Brownies

Born in Edinburgh in 1850, Robert Louis Stevenson wrote several highly acclaimed books, including *Treasure Island*, *The Strange Case of Dr Jekyll and Mr Hyde*, and *Kidnapped*. Stevenson was fascinated by sleep and dreaming throughout the whole of his life. As a child the great novelist frequently experienced night terrors, and as a young man he would enjoy vivid dreams involving fantastical adventures in far-away places. As time went by Stevenson realized that he often experienced complete stories throughout the night, and was even able to return to the same story on successive nights and witness different endings. Eventually, he managed to train himself to remember his dreams and used them as a basis for several of his plots.

Later in life, Stevenson described this process in an extraordinary essay entitled 'A Chapter on Dreams'.[26] He noted that dreams were like a '. . . small theatre of the brain which we keep brightly lighted all night long'. His own nocturnal theatre was populated by a cast of 'little people' that he referred to as his 'Brownies'. As time went by, Stevenson's imaginary cast started to produce plays that acted as the basis for many of his novels. Perhaps the most famous instance of this Brownie-based brainstorming provided the inspiration for his most celebrated story, *The Strange Case of Dr Jekyll and Mr Hyde*.

Stevenson had spent several days trying to find a suitable plot for his new book, and eventually decided to call on the help of his beloved Brownies. After a few hours asleep he drifted into a dream and found himself watching a scene in which a good man was being pursued for some type of crime. In a panic, the character ingested a kind of powder or potion, and transformed into the personification of evil. At that moment Stevenson screamed out in his sleep, and his wife woke him up. The great novelist was less than delighted, claiming that he was dreaming of a 'fine bogey tale' and had now, quite literally, lost the plot. However, despite being unexpectedly pulled away from the Brownies' performance, Stevenson was able to complete his most famous work of fiction in just a few days.

This inspirational use of dreaming eventually became an essential part of Stevenson's work, with the great writer noting, 'When I lay down to prepare myself for sleep, I no longer seek amusement, but printable and profitable tales.' Towards the end of his essay Stevenson explains that he doesn't fully understand this mysterious process ('Who are they then? My Brownies, who do one-half my work for me while I am fast asleep?'), but says that he had noticed that it often takes a threatening letter from his bank to kick-start the Brownies into action.

Another great Scottish writer, Sir Walter Scott, also relied on a similar process and once noted: 'When I have got over any knotty difficulty in a story . . . it was always when I first opened my eyes that the desired ideas thronged upon me. This is so much the case that when I am at a loss, I say to myself, "Never mind, we shall have it at seven o'clock tomorrow morning." '[27]

Stevenson and Scott are far from the only writers to have found inspiration in their dreams. In 1797, for instance, Samuel Taylor Coleridge fell asleep one afternoon and dreamt over 200 lines of a new poem. When Coleridge woke he furiously scribbled down the lines, but was interrupted by a visitor. After the visitor had left Coleridge was unable to remember the last lines he had composed during his sleep, with the consequence that his best-known poem, 'Kubla Khan', remained unfinished.

More recently, horror writer Stephen King has described how the plots for many of his books have come to him in his dreams. For example, the genesis of his bestselling book *Misery*. In the early 1980s, King fell asleep while on a flight and dreamt about an author who was held captive by a psychotic fan. King woke up, and scribbled down some notes on an airline napkin. The following night he started to write his now-classic novel.

Other works of fiction inspired by dreams include Mary Shelley's *Frankenstein*, Charlotte Brontë's *Jane Eyre*, and Stephenie Meyer's *Twilight* series (evidence that, in my opinion, even the Brownies have off days).

Such dream-like inspiration is not limited to writers. In fact, many musicians have created their best work while sound asleep. For instance, the eighteenth-century Italian baroque composer

Giuseppe Tartini dreamt that he had sold his soul to the devil, woke up, and produced his now famous 'Devil's Trill Sonata'. Several centuries later the same dream-like inspiration helped Paul McCartney to create one of The Beatles' most famous songs. In May 1965, McCartney was staying with his mother in London. One morning he woke up with the tune to 'Yesterday' fully formed in his mind, but was convinced that he couldn't have composed it in his sleep. After playing it to several colleagues, McCartney slowly realized that it was his own creation and started to work on some lyrics. After trying with various options McCartney settled on lyrics that referred to the death of his father, and created his classic song.

Similarly, many athletes have changed the face of sport in their dreams. The world-famous golfer Jack Nicklaus had a dream in which he was holding his club in a different type of grip. The following day Nicklaus tried out the new grip for real and suddenly experienced a run of success. Heavyweight boxing champion Floyd Patterson would create new types of punches in his dreams, and then try them out for real in the ring. And the world-class acrobat Tito Gaona described how she would dream about new moves and routines when she was sound asleep, and then re-create them the following day.

The same remarkable nocturnal process has resulted in several scientific and technological discoveries that have changed the course of history.

Chemist Dmitri Mendeleev created the modern-day periodic table after a dream in which he imagined the elements being related to one other in the same way that musical notes are part of a larger structure. Similarly, chemist August Kekulé figured out the structure of the chemical compound benzene, after dreaming about a snake biting its own tail and realizing that the elusive compound was composed of a ring of carbon atoms.

At the turn of the last century the Indian mathematical genius Srinivasa Ramanujan teamed up with Cambridge scholar Godfrey Hardy and made a series of breakthroughs in number theory. Ramanujan was often inspired by his dreams, claiming that the Hindu goddess Namagiri would frequently appear to him during the night and present him with formulae. Around the same time, the German pharmacologist Otto Loewi dreamt about how to carry out an experiment that would reveal how nervous impulses travelled around the body. Loewi's experiment formed the basis for work that eventually led to him receiving a Nobel Prize and being labelled the 'father of neuroscience'.

Dreams have also played a key role in several inventions. In the 1840s Elias Howe created the first sewing machine after dreaming about spears with holes near their tips, and realizing that placing a hole at the tip of the needle would prevent the thread from catching after it went through cloth. More recently, Harvard electrical engineer Paul Horowitz has described how he regularly relied on his dreams when he was trying to design radio telescopes. In his dreams, Horowitz would watch a man working on the same problem that he himself faced in real life. Often the mythical man in Horowitz's dream would come up with a solution.

Given the inspirational role that dreams have played in the arts and science, it is perhaps no surprise that the French Symbolist poet Saint-Pol-Roux used to hang a sign on his bedroom door proclaiming 'Poet at work', and author John Steinbeck attributed dream-based breakthroughs to the 'the committee of sleep'.

The secret science of the bedtime story

Parents have read bedtime stories to their children for generations. These stories allow parents to spend time with their children, help promote their language skills, and gently help them fall asleep. But could they also serve another vitally important purpose? During dreaming, your unconscious mind works through any problems that you are facing in your life and, like a skilled therapist, tries to soften the emotional impact of these issues and find novel solutions. The bedtime stories that we tell our children seem perfectly designed to complement this process. We read these stories to children just as they are falling asleep, and many of the classic tales contain important life lessons that help them cope with everyday stresses and concerns. 'The Boy Who Cried Wolf' underlines the importance of not telling lies, 'The Ugly Duckling' encourages children not to judge people by their looks, and 'The Three Little Pigs' is all about the power of perseverance.

Could it be that the structure and timing of bedtime stories have evolved to give young minds more pleasant and productive dreams? And, if this is the case, is it possible to create even more effective stories that are explicitly designed to reflect the psychological principles that underlie a happy, productive, and resilient life? To find out, I teamed up with educator and writer Jenny Hambelton to create the world's first science-based bedtime story.

We decided to focus on an important aspect of resilience that helps give people hope when bad things happen to them. The idea forms the basis of a well-known Zen parable concerning a wise farmer. The parable starts with the farmer waking up one morning and discovering that his best horse has run away. The farmer's neighbours hear the news and say, 'What bad luck!' However, the wise farmer simply replies, 'Maybe.' That evening the farmer's horse comes back and brings a wild horse with it. Upon hearing this news, the farmer's neighbours exclaim, 'What wonderful luck!' and again the farmer replies, 'Maybe.' The next morning, the farmer's son climbs on the untamed horse, is thrown off, and breaks his leg. The farmer's neighbours see this and say, 'What bad luck!' Again, the farmer replies, 'Maybe.' Finally, that evening, some military officers arrive at the village to draft the young men into the army, but have to leave the farmer's son because of his broken leg. As a result, the neighbours congratulate the farmer on his amazing good luck.

This simple parable illustrates the idea that seemingly negative events can turn out to be surprisingly positive in the long run, and so it's important not to feel too down-hearted when bad things happen. We took this simple notion as our starting point, and created a story that children would understand, identify with, and enjoy. The resulting story is called 'Donald and the Elephant at School' and is suitable for children aged four to six (see Appendix). In the story a young boy called Donald finds a friendly elephant on the way to school. Donald befriends the elephant and takes him into classes. Unfortunately, things do not start well. The elephant throws paint around, eats everyone's food, and ruins a music lesson. Donald thinks that finding the elephant was a disaster, but then the elephant saves the day during a school trip and Donald realizes that sometimes everything works out for the best.

The language in the story has been carefully chosen to maximize the likelihood of key elements influencing a child's dreams. Psychologists have long known that words that are easy to imagine (such as 'chair' and 'house') are also much easier to remember than those that do not instantly conjure up a mental image (such as 'peace' and 'justice').[28] During this work, researchers have presented people with thousands of words and asked them to rate how easy it is to imagine each word.[29] When we wrote 'Donald and the Elephant at School' we used these lists to identify words that are extremely easy to imagine (such as 'elephant', 'paint', and 'water') and built the story around these words. In addition, other research has shown that strange and unusual combinations of images (such as a large mouse chasing a small cat) are more memorable than mundane combinations (such as a large cat chasing a small mouse). So we designed the story to incorporate a series of unusual scenes, such as an elephant throwing paint around, jumping in a swimming pool, and trumpeting loudly during a music lesson. Finally, the story ends with Donald wondering whether the elephant will come back to school the next day, but never seeing him again. This final line is designed to provoke a sense of curiosity, encourage children to think about where the elephant went and what happened next, and so increase the likelihood of the story affecting their dreams. We have also included a few simple comments for parents to make after the story in order to help cement its meaning in place.

If you have children, read the story to them at bedtime. Then, the following morning, ask them to describe any dreams that they had during the night and see whether the positive theme of the story has had a beneficial effect on their sleeping mind.

> The science of sleep has provided a remarkable insight
> into the importance of bedtime stories. Armed with this new
> perspective, I believe that it's possible to create stories that
> boost children's psychological health during the night and so
> help ensure that everyone really does live happily ever after.

In this lesson we have uncovered the surprising scientific truth
about dreaming. Each night your dreams act like a 'nocturnal
therapist' who helps you to identify concerns in your waking life,
and tries to find innovative solutions to your problems. Over the
years these therapists have worked away night after night, helping
millions of people, and inspiring countless writers, musicians, and
scientists. Perhaps the most exciting aspect of the research shows
that it's possible to make the most of your 'nocturnal therapist'
by learning a few simple dream-based techniques. Dreams are
far from the meaningless froth on the beer of sleep.* In fact, they
have the power to improve your life, and even change the world.

---

* Sigmund Freud first used the term 'dreams as froth' (*träume sind schäume*) to
describe theories suggesting that dreams are essentially meaningless.

ASSIGNMENT

Experimenting with your eyes and arms

Before we begin our final Night School lesson, I'd like you to carry out two simple experiments involving your eyes and arms.

The eyes experiment

For this experiment you will need ready access to either a friend or a camera. Let's start off by assuming you have a friend around. Ask them to stand in front of you. No, not that close. About two feet away. That's better. Next, keep your head facing forwards, and then roll your eyes up towards the ceiling as far as they will go. All being well, you will be looking a few feet above your friend's head. Now, keep looking up, and try to close your eyelids. When you do this, you might be tempted to lower your eyes. Fight the temptation. Finally, ask your friend to quickly look at the amount of white showing between the top of your lower eyelid and the base of your cornea, and then use the chart overleaf to convert this observation into a number between zero and four. The experiment should only take a moment, and please do not continue with it if you experience any discomfort.

If a friend is not around, simply hold a camera in front of your face, carry out the procedure described above, and take a snapshot. You can then see where you sit on the scale by looking at the resulting photograph.

Please make a note of your score overleaf.

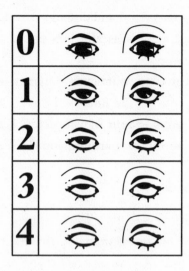

Score: _____

The arms experiment

As you might imagine, this experiment involves your arms. Hold out both of your arms directly in front of you, ensuring that they are parallel to the floor and that your two hands are level with one another. To the uninformed observer, it will appear as if you are mid-sleepwalk.

In a few minutes I am going to ask you to close your eyes for about thirty seconds. When you do, I want you to imagine that there is a helium-filled balloon attached to the fingers of your right hand, and a pile of heavy books attached to your left hand. The helium balloon will be pulling your right hand high into the air, and the books will be pulling your left hand towards the ground. Do not consciously move your hands, but try to imagine the balloon causing your right hand to drift upwards and the weight of the books dragging your left hand downwards. After around thirty seconds or so, open your eyes and look at the position of your hands. OK, away you go.

Thanks. You can put your arms down now. Please use the following scale to convert the final position of your hands into a number between zero and four (if your left hand was higher than your right hand, stop messing around and repeat the test).

Score	Description
0	Your right and left hands are perfectly level.
1	Your right hand is less than an inch above your left hand.
2	Your right hand is between about one and three inches above your left hand.
3	Your right hand is between three and six inches above your left hand.
4	Your right hand is more than six inches above your left hand.

Please write your score below.

Score: _____

Finally, please add up your scores from the two experiments, and make a note of your total score below.

Total score: _____

Many thanks.

When I first started working at the University of Hertfordshire I had an office on the top floor of the psychology department. In addition to all of the furniture that you would usually expect to find in a university office, the room housed a large black reclining chair. I soon discovered that the chair was extremely comfortable and, from time to time, would use it to take a quick nap. At the time, I had no idea that over the years the chair had played host to thousands of similarly sleepy people.

The university's psychology department was founded in the early 1970s by an eccentric academic named Tony Gibson. Gibson led a colourful life that included playing a key role in the British anarchist movement, spending some of the Second World War in prison as an unregistered conscientious objector, and appearing as the 'Brylcreem Boy' in a series of national hair gel advertisements. I never had the pleasure of meeting Gibson, but by all accounts he tended to be both abrupt and outspoken, with a colleague once noting: 'I wish there were more people like Tony. But not *too* many.'[30]

Gibson carried out several groundbreaking studies into hypnosis, with much of his work taking place in the big black chair that eventually found its way into my office. Over the years Gibson investigated various fascinating topics, including the relationship between hypnosis and dreaming.[31]

Many people think that hypnosis is a form of sleep. In fact, nothing could be further from the truth. When people are hypnotized they can usually hear perfectly well, and have no problem responding to a hypnotist's detailed suggestions. In contrast, when people are asleep they are much more isolated from the outside world, and can rarely carry out complex behaviours. For these reasons, most psychologists see the hypnotic state as closer to being awake than asleep. But does that mean there is no relationship between hypnosis and dreaming?

Both the eye-rolling and arm-raising experiments are reliable tests of how susceptible you are to being hypnotized.[32] The higher your final score, the more easily you can be hypnotized, with a score of six or more indicating that you are highly hypnotizable (suggesting that at some point in your life you will find yourself onstage tucking into a tasty onion).

Gibson found several major differences in the dreams of people who are, and aren't, especially susceptible to hypnosis. Highly hypnotizable people experience dreams that are especially vivid, colourful, pleasant, bizarre, and exciting. In addition, they are more likely to use their dreams as a source of creative inspiration and find meaning in them. Given that this is the case, it perhaps isn't surprising that they also tend to remember more of their dreams, and believe that they contain paranormal clues about future events. So, the next time you want to find out what really goes through someone's mind at night, ask them to roll their eyes and raise their arms.

Gibson also discovered that highly hypnotizable people are often able to take control of their dreams, and so experience their perfect nocturnal fantasies. However, the good news is that you don't have to be an extreme arm-raiser in order to enjoy this wonderful experience. For the past hundred years, sleep scientists have developed several techniques to help everyone take control of their dreams. Best of all, we will be exploring these techniques in our final Night School lesson.

Lesson 8

SWEET DREAMS

Where we find out how to control your dreams,
banish nightmares, and have a lucid dream.

Welcome to your final lesson at Night School. In the previous seven lessons we have travelled deep into the night, and discovered how to radically improve your sleep and use your dreams to change your life. In this final lesson we are going to take a close look at perhaps the strangest aspect of sleep science.

In the science fiction film *Inception*, a character named Dominic Cobb is a 'dream architect' who enters sleepers' subconscious minds and manipulates their dreams. *Inception* proved to be one of the highest grossing films of all time, and the idea of dream control has caught the public imagination. It's easy to see why. After all, controlling your dreaming mind would allow you to experience any fantasy you desire: from spending quality time with your favourite celebrity to staging a thrilling bank heist, flying through space, or lying on a sun-kissed beach – your imagination would be your only limit.

For more than a century, sleep scientists have experimented with many different ways of trying to control dreams. Some of these techniques proved to be highly effective and others were a complete failure. In this lesson we are going to explore these ideas, and sort the pseudo-scientific chaff from the evidence-based wheat. Along the way we are going to discover how to induce a lucid dream, find out the results from the world's largest dream control experiment and learn how to banish nightmares.

We begin our journey by meeting a man who has devoted his life to living the dream.

On becoming lucid

As a child, Stephen LaBerge enjoyed going to the cinema each week and watching his matinee idols act out the latest instalment of a thrilling adventure. One night he had an exciting dream about being an underwater pirate and, inspired by his heroes of the silver screen, he wondered whether he would be able to persuade his sleeping brain to return to his imaginary adventure the following night. LaBerge found it surprisingly easy to re-enter the dream, and spent the next few weeks repeatedly returning to his aquatic fantasy. During these dreams, LaBerge gradually become aware that he was able to hold his breath while underwater for a surprisingly long period of time, and this simple realization eventually resulted in him feeling as if he were awake during his pirate adventures. LaBerge had experienced a strange phenomenon that sleep scientists now refer to as 'lucid dreaming'. During these fascinating episodes, people are aware that they are dreaming and, on a good night, can even take control of their fantasy world.

LaBerge's brief spell as an underwater pirate kicked off a life-long interest in lucid dreaming. In the late 1970s, he decided to explore the topic in a more systematic way, and enrolled for a doctorate at Stanford University. At the time, many sleep scientists were somewhat sceptical about the existence of the phenomenon, arguing that it was impossible for the sleeping mind to become conscious or to take control of a dream. If these odd experiences were happening at all, argued the sceptics, then they must be occurring during micro-awakenings rather than during dreaming. LaBerge set out to prove the sceptics wrong.[1]

He decided to wire himself up to an EEG machine, fall asleep, have a lucid dream, and then analyse his brainwaves to discover whether he had been in REM. There was, however, one small problem with his plan. In order to know which part of the EEG data to examine, he needed to be able to indicate when he was having a lucid dream. Unfortunately, LaBerge was well aware that his body would be almost completely paralysed when he was dreaming, and so making any kind of signal was going to be a tad tricky. After much head scratching, he hit on a clever idea. Research had shown that during a dream, men's penises and eyes remained active, and so it might be possible to use either of these two moving parts to send a sign. Thankfully, LaBerge opted for trying to move his eyes.

In January 1978, LaBerge conducted the first in a series of now-classic experiments. He connected several EEG sensors to his head, climbed into bed, and settled down for the night. After a few hours of sleep, he opened his eyes and looked around the room. Everything appeared fine, except that he was unable to feel or hear anything. Realizing that he was probably having a lucid dream, he took command of his imaginary world and slowly waved his hand in front of his face. He then tried to track his imaginary hand by actually moving his eyes from side to side. A few seconds later the experience faded and he woke up.

When LaBerge examined his EEG data he was thrilled to see that the sensors had detected several very large eye movements. As a result, he was able to pinpoint the exact moment that he had been having a lucid dream. He looked at his brain activity from this part of the night, and was delighted to discover that this lucid dream had indeed taken placed during REM. LaBerge had proven the sceptics wrong.

LaBerge has spent much of his subsequent career developing various psychological techniques that help people to have lucid dreams, including imagining their perfect dream before they fall asleep, and getting into the habit of regularly conducting various 'reality tests' to discover if they are in a lucid dream or not (see opposite). He has also created several gadgets to help would-be lucid dreamers, including a high-tech blindfold that monitors their eye movements throughout the night and that activates several bright lights in its centre whenever it detects a significant amount of eye movement. All being well, the person 'sees' these lights in their dream, and this helps them become aware that they are dreaming.

These techniques have allowed thousands of people across the world to create their perfect dream. Perhaps predictably, many of them have used the opportunity to engage in the ultimate form of safe sex. Curious about this somewhat seedy side of the night, LaBerge once invited an experienced female lucid dreamer to his laboratory and monitored her brain activity, eye movements, and vaginal activity. The volunteer was asked to try to have a sexy lucid dream, and then move her eyes in a predetermined way when she reached orgasm. During the night the woman's eye movements suggested that she was having the best of times for one fifteen-second period. Throughout this time her vaginal activity was at a maximum, suggesting that lucid dreams do indeed have the power to stimulate the genitalia and create genuine sexual experiences.

When they are not exploring people's dream orgasms, researchers have delved deep into the strange science of lucid dreaming, and examined whether this weird phenomenon can be used to help solve more down-to-earth problems.

A beginner's guide to lucidity

Stephen LaBerge and several other sleep scientists have developed a variety of psychological techniques to help you gain conscious control of your dreams. In 2012, Tadas Stumbrys from Heidelberg University reviewed these techniques to help determine which had the highest success rate.[2] Some of them are carried out during the day while others are intended for the night. Here is a brief summary of the most effective procedures. Feel free to mix and match whichever ones seem to work for you.

During the day

Making time: Look at your watch between five and ten times each day, and check that you can see the numerals properly each time. After a week or so, this strange 'watch-looking' procedure will become a habit, and you will start to do it in your dreams. During a dream you will struggle to read the numerals on your watch dial, and this will help you to become lucid.

Visualize your perfect lucid dream: Find a few minutes each day to lie down, close your eyes, and imagine how it would feel to have a lucid dream. Visualize who you would most like to meet in your dream and what you would love to happen. Enjoy the experience and allow your imagination to create your wildest fantasy.

When you go to bed

Set your intention: Place a pen and pad on your bedside table. Just before you go to bed, tell yourself that you want to wake up towards the end of a dream, and be able to remember what thoughts and images were going through your head. If you are successful, write down a brief description of your dream. Then, as you drift back to sleep, imagine yourself returning to the dream, but becoming lucid. Visualize yourself back in the same scenario, but doing something that will reveal that you are in a dream. There are certain experiences that your brain struggles to produce during a lucid dream (perhaps because they require too much thinking power, or because you have never experienced them in real life), and so you might, for example, look in a mirror (lucid dreamers tend to see a blurred reflection), levitate off the ground, turn on the lights (in a lucid dream it isn't possible to alter the illumination levels), or check the numerals on your watch.

Raise the alarm: Set your alarm clock to wake you up about an hour before you normally get out of bed. When you wake up, spend the next thirty minutes reading a book, working on a jigsaw puzzle, or quietly writing about the thoughts that are going through your head. Now get back to sleep. This technique, known as 'sleep interruption', increases the likelihood of a lucid dream about twentyfold.

When you have a lucid dream

Take it easy: Many first-time lucid dreamers become very excited once they are aware that they are dreaming, and this feeling of excitement terminates the dream. Try to stay relaxed and calm.

Go for a spin: If you think that the lucid dream is about to end, then try imagining rubbing your hands together, or spinning around like a top. According to experienced lucid dreamers, this sense of movement will help you remain in the dream.

Finally, if all of these techniques fail, try playing computer games. Research shows that those who regularly play first-person games have more lucid dreams than non-gamers.[3]

Unicycling, snowboarding, and skateboarding

Research into lucid dreaming has revealed that around 50 per cent of adults have had this strange experience at least once in their lives, and that one in five people experience it on a monthly basis.[4] Unlike many other psychological phenomena, there is no relationship between the frequency of lucid dreaming and a person's age, sex, or personality. By asking lucid dreamers to move their eyes whenever the experience started and stopped, researchers were also able to discover that most lucid dreams last about two minutes, and tend to take place early in the morning. Perhaps most intriguing of all, scientists have uncovered several surprising differences between lucid dreams and the more run-of-the-mill experiences that mere mortals have on a nightly basis. For example, the curious work on 'dream characters'.

As we discovered in Lesson 7, most dreams are populated by the dreamer's friends, family members, and work colleagues. In contrast, lucid dreams frequently feature complete strangers. Discovering more about these mysterious characters isn't easy, in part because lucid dreamers often report that staring at these strangers causes them to cover their face, turn off the lights, and wear a hood in future dreams.[5] However, some lucid dreamers have slowly gained the trust of these shy, imaginary individuals, and eventually persuaded them to take part in several tests.

In the late 1980s, a pioneering German sleep scientist named

Paul Tholey told a group of lucid dreamers to ask their dream characters to make a drawing and name a word that wasn't known to the dreamer. Even through the lucid dreamers were unaware of each other's reports, remarkably similar results emerged. The dream characters were perfectly capable of making quite impressive drawings, with one of them even showing off by creating a strange-looking sketch and then turning it upside down to reveal that it was a self-portrait. Even more amazingly, some of the characters were also able to name a word that was not known to the lucid dreamer. For example, one male lucid dreamer bumped into an imaginary female friend, and asked her to name a foreign word that he was unfamiliar with. The woman remarked 'Orlog', and went on to explain that the word described their relationship. In the morning the lucid dreamer looked up the word and discovered that it was Dutch for 'quarrel'.

However, other work suggests that the dream characters don't have much of a head for numbers. In several studies, researchers have asked lucid dreamers to set their dream characters increasingly difficult mathematical problems.[6] This has proved a tad tricky because many of the dream characters were reluctant to participate. In one instance, for example, a lucid dreamer asked one imaginary man if he would mind doing some arithmetic problems. The man agreed, and so the lucid dreamer asked him to calculate 4 times 4. The man confidently replied '16', and requested a harder problem. The lucid dreamer decided to up the ante by asking him to calculate 21 times 21. Suddenly, the man waved goodbye and promptly disappeared. However, several lucid dreamers have managed to persuade their dream characters to sit a series of increasingly difficult tests. The results revealed that the characters were fine with very simple sums, but struggled when the answer was greater than 20. The research also suggests that the types of maths-based stereotypes that exist in the real world also hold true in lucid dreams, with male dream characters exhibiting better mathematical skills than their imaginary female counterparts.

Further work into lucid dreaming has examined whether the technique can help improve people's lives. In some studies, researchers have asked experienced lucid dreamers to practise a physical skill during their dreams, and then assess the effect of these nocturnal fantasies on the dreamer's ability to perform these skills in the real world. Once again, much of this work was carried out by Paul Tholey. As well as being a highly respected academic, Tholey was also a top-notch lucid dreamer, skateboarder, snowboard fanatic, and cyclist. He decided to experiment on himself and discover whether lucid dreaming could help him hone his skills. After much trial and error, Tholey claimed that his nocturnal fantasies had helped him to balance blindfolded on a unicycle, snowboard without bindings, and perform a handstand on a skateboard. Although his claims were greeted with more than a pinch of scepticism from many mainstream scientists, recent research suggests that the skateboarding academic was on to something.

In 2010, Daniel Erlacher, now at the University of Bern in Switzerland, decided to carry out the first laboratory-based experiment to examine whether lucid dreamers can enhance a physical skill. Erlacher asked a group of lucid dreamers to practise throwing a coin into a cup in their dreams, and then tested their real-life coin-throwing skills the next morning.[7] As predicted by Tholey's experiences with unicycles and skateboards, the lucid dreamers who reported being able to practise when they were sound asleep were better at the task.

On the upside, all of this imaginary exercise does not make people feel tired when they wake up. On the downside, learning to become conscious during a dream is very tricky and some people never manage it. Luckily, for more than a century a handful of researchers have experimented with a much easier way of taking control of the night. Their story starts with one of the very first sleep scientists.

Sweet dreams are made of cheese

In Charles Dickens's *A Christmas Carol*, Ebenezer Scrooge tries to convince himself that the ghost of Jacob Marley is nothing more than a crumb of undigested cheese, and famously declares that 'There's more of gravy than of grave about you.' This passage helped give rise to the popular myth that eating cheese affects your dreams, but is there any truth behind Scrooge's panicked utterances?

To find out, in 2005 the British Cheese Board asked 200 volunteers to spend a week eating a small piece of cheese before going to sleep and then report their dreams. This fun experiment disproved the notion that cheese gives you nightmares, with the majority of the volunteers enjoying a pleasant night. Interestingly, the British Cheese Board also announced that the type of cheese consumed seemed to affect the volunteers' dreams. Stilton caused bizarre dreams (with one Stilton eater dreaming about soldiers fighting each other with kittens instead of guns), Cheddar eaters dream about celebrities, Red Leicester causes people to dream about their past, and one Brie eater dreamt about having a drunken conversation with his dog.

Alternatively, if you want to have an erotic dream, bury your face in the pillow. Psychologist Calvin Kai-Ching Yu, from Hong Kong Shue Yan University, has examined the relationship between sleeping position and dreaming. Interviews with almost 700 people revealed that sleeping on the stomach is strongly associated with sexual dreams, having a love affair with a celebrity, and dreams in which you appear naked (or, on a really good night, all three).

A similar study, conducted by researchers in Turkey, has revealed that, compared to those sleeping on their left side, right side sleepers tend to have more dreams that are associated with happiness and hope, enjoy better sleep quality, and experience significantly fewer nightmares.[8]

Dreams and How to Guide Them: Practical Observations

The Marquis d'Hervey de Saint-Denys was a kind of nineteenth-century Hong Kong Phooey. By day, Hervey de Saint-Denys lived the life of a mild-mannered French academic with a scholarly interest in Chinese culture and customs. But each night he threw off his academic gown, donned a nightcap, and delved deep into the science of dreaming.

Born in 1822, Hervey (as he was known to his friends) started to record his dreams when he was just thirteen. He soon discovered that he was able to remember at least one dream each night, and became obsessed with recording these nocturnal adventures in as much detail as possible. He spent the next twenty years describing his dreams, eventually filling twenty-two notebooks with his nightly reports and illustrations. Tragically, these notebooks have become lost in the mists of time, and so modern-day sleep scientists are not able to benefit from Hervey's nightly reports. However, on the upside, the great French dreamer did leave a lasting account of some of his experiences in his book, *Les Rêves et les Moyens de les Diriger; Observations Practiques* (or, as I tend to refer to it, 'Dreams and How to Guide Them: Practical Observations').

First published anonymously in 1867, this now-classic text describes a series of fascinating dream-related thoughts and investigations. A large part of the book is given over to Hervey's experimental work into lucid dreaming. In one chapter, for instance, he notes that if dreams are based on waking experiences then it shouldn't be possible to dream about something that you have never experienced. Eager to put this idea to the ultimate test, Hervey tried his best to kill himself during several lucid dreams. Each time he failed to dream about his untimely demise, and so concluded that waking experiences do act as the building blocks for our nightly excursions.

Some of Hervey's most innovative work focused on whether it was possible to influence his dreams. He decided that the best way forward involved a simple two-stage procedure. First, he would spend his waking hours enjoying some kind of pleasant experience while sniffing an unusual smell. Then, when he was asleep, he would expose himself to the same smell, in the hope that it would cause him to dream about the experience.

To test the technique, he purchased a bottle containing an unusual perfume, and spent two weeks with a group of friends in the mountains. Whenever he came across a particularly striking mountain view, Hervey whipped out the perfume bottle, placed some of the scent on his handkerchief, and sniffed away. When he returned home, he arranged for his servant to quietly creep into his bedroom at randomly determined times, and pour a small amount of the scent on his pillow. The experiment was a huge hit, and resulted in Hervey enjoying several joyous mountain-related dreams.

Thrilled with his success, he decided to repeat the study using sound instead of scent. He attended several balls, and arranged for the orchestra leader to play a particular piece of music whenever he danced with an especially attractive woman. Hervey then built a special music box that played the same tune, and constructed a mechanism to link the operation of the box to his bedside clock. At predetermined times throughout the night the music box suddenly jumped into action and played the selected tune. Once again the study was a spectacular success, with the night music causing Hervey to enjoy several erotically charged dreams.

Hervey's technique, although fascinating and effective, suffered from one major problem. Although almost everyone would like to be able to have sweet dreams on demand, most people are not prepared to spend their days associating particular experiences with unusual smells and sounds. Would it be possible, wondered researchers, to obtain the same effects by exposing dreamers to subtle stimuli that already had strong associations? Would, for instance, people's dreams become more pleasant if you secretly sprayed some sweet-smelling rose scent in their bedrooms, or take a turn for the worse if you gently whispered death threats into their ear?

Several anecdotal reports suggested that there might be something to this seemingly far-fetched idea. Take, for example, the remarkable experience reported by a well-known British general and his wife.[9] Towards the middle of the nineteenth century, General Sir William Sleeman was sent to India to help suppress a violent secret society known as the 'Thuggee' (which, incidentally, is the origin of the word 'thug'). One night, Sleeman and his wife pitched their tent in a field and went to bed. Soon after falling asleep, Sleeman's wife had a horrible dream about corpses, woke up, and asked her husband to move their tent to another location. When General Sleeman walked outside the tent he noticed a very weak, but somewhat unpleasant, smell. Curious, the good General dug up the ground, uncovered the decomposed bodies of fourteen victims of the Thuggee, and concluded that the faint smell of death had influenced his wife's dreams.

Although compelling, such reports could always be dismissed as coincidence, and so researchers set about conducting more controlled investigations. Many of these studies were carried out by the nineteenth-century French psychologist, Alfred Maury. Over the course of several months, Maury arranged for his friends to subject him to several strange stimuli when he was sound asleep, including sticking a straw between his toes and stroking his face with a feather. Time and again the stimuli influenced Maury's dreams, with, for instance, the straw causing him to dream about having a stake driven through his feet, and the feather forming the basis of a terrible torture-based dream in which he had hot pitch poured onto his face.

Other researchers eliminated the need for their colleagues to creep around in the dark by using the cutting-edge technology of the day to deliver the stimuli.

In 1899, an American physiologist named Leonard Corning proudly unveiled the most elaborate of these devices (see diagram). Those wishing to try out Corning's cutting edge 'dream machine' were invited to lie down on the couch and have a leather helmet firmly secured on their head. The helmet resembled the type of head gear now used by amateur boxers, and was designed to hold metallic saucers over the volunteer's ears. Twenty-five-feet-long lengths of rubber tubing were then used to connect the saucers to a nearby Edison phonograph, and during the night Corning piped various pieces of music into the volunteer's ears ('. . . harmony is more effective than melody . . . and for this reason selections from the Wagnerian compositions render excellent service').

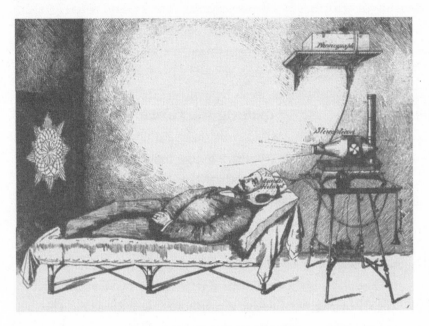

Leonard Corning's dream machine.

Corning claimed that his remarkable device could help those suffering from nightmares, and described a case in which one man's 'carnival of the horrible' was transformed into 'agreeable visions' over the course of just a handful of sessions. In another instance, the procedure helped a woman suffering from depression, although the lady in question also exhibited 'a striking gain in appetite' that culminated in 'a small though decided accession of weight'.

Although undoubtedly years ahead of its time, the rather limited technology available to Corning meant that his remarkable invention failed to catch the public imagination. Like all of the late nineteenth-century dream architects, Corning wasn't sure when his volunteers were dreaming. As a result, trying to play a sound or spray a scent at the right time was very much a hit-and-miss affair. Faced with this seemingly insurmountable problem, researchers turned their back on dream control and didn't give the idea a second look for more than fifty years.

Smelling the roses

In the nineteenth century, several pioneering researchers examined the effect of smell on dreams. Unfortunately, this type of research fell out of favour for many years, and has only recently been revived.

In 2009, Michael Schredl, from the Central Institute for Mental Health in Mannheim, examined the relationship.[10] Schredl and his team arranged for a group of volunteers to experience one of two smells while they were dreaming. One of the smells was very pleasant and reminded most people of freshly cut roses. The other smelt of rotten eggs. The following morning, the volunteers were asked to describe their dreams, and the researchers rated the subsequent reports for positive and negative emotion. Even though they were completely unaware of the smells, the volunteers who had been exposed to the more pleasant smell had much more positive dreams.

So, if you want to have sweet dreams, ensure that your bedroom contains just a hint of your favourite smell.

Dream:ON

In Lesson 1, I described how Eugene Aserinsky revolutionized sleep science by discovering the association between dreaming and 'rapid eye movement' (REM). For the first time in history, sleep scientists were able to pinpoint when a person was dreaming, and several researchers realized that this discovery had huge implications for dream control. Instead of spending the night randomly playing subtle sounds and hoping for the best, researchers now knew exactly when to press play.

Much of the early work into dream control was carried out in the early 1960s by Stanford sleep scientist William Dement. Dement invited volunteers into his sleep laboratory, waited until they were dreaming, and then secretly played a tone into their ears, shone a bright light onto their face, or sprayed them with water. Dement then waited ten minutes, woke up the participant with the aid of a loud dinner bell, and had them describe their dream. About half of the participants incorporated the stimuli, with, for instance, the water resulting in dreams about sudden rainfall, the tone causing people to dream about an explosion, and the bright light resulting in reports of the sudden outbreak of fire. Compiling a league table of stimulation, Dement reported that the light was only incorporated into 9 per cent of the dreams, the tone into 23 per cent and the water into a massive 42 per cent.

In an additional set of studies, Dement played tape recordings of various sounds to his sleeping subjects, including a police siren, a steam locomotive, a squeaky door, and a speech by Martin Luther King (presumably 'I have a dream . . . and right at this moment, so do you'). Once again, many of these sounds were incorporated into people's dreams. For instance, after hearing the Martin Luther King speech, one volunteer reported a dream in which they were eating in a dormitory and suddenly saw a man stand up and talk about racial tension. Later on in the dream, the same imaginary man emphasized the need for people to love one another. Around half of the sounds were incorporated into participants' dreams, with police sirens and steam trains being especially effective, and squeaky doors and crickets chirping coming bottom of the class.

In the late 1960s, psychologists built on this work by playing people's names into the ears of dreamers, and discovered that about half of the names subsequently made an appearance in their dreams. Playing dreamers a recording of their own name made the figures more assertive and independent, whereas the voice of a stranger resulted in a character that tended to be passive and unhelpful.[11]

After this initial flurry of research, the idea once again dropped by the wayside. Although researchers had discovered when best to stimulate dreamers, and which types of signals worked well, they had no way of exporting these techniques beyond their sleep laboratories. I recently wondered whether it might it be possible to use this work as the basis of an iPhone app that could help millions of people create their perfect dream. I contacted YUZA, one of Britain's top iPhone app development companies, and suggested that we conduct a large-scale project into dream control. A few weeks later I met up with company's CEO, and they kindly agreed to come onboard. And so the adventure began.

The company assembled a team of top developers and started to put together the app. It was an ambitious project that pushed the iPhone to the very edge of its capability. Slowly but surely, however, the 'Dream:ON' project took shape. The idea is simple. Before going to sleep, people indicate on their iPhones what time they would like to wake, and select a specially prepared 'soundscape', such as a walk in the countryside (think breeze rustling through the trees) or a visit to the coast (waves gently lapping at the shoreline). They then place their iPhone onto their mattress and go to sleep. The app works by targeting the final REM period in the sleep cycle. About thirty minutes before the person wants to wake up, the iPhone's movement sensors jump into action. When the sensors detect the lack of movement associated with dreaming, the app gently plays their chosen soundscape that then, in theory, influences their dreams. When they wake up the person is then prompted to submit a description of their dream, and all of these anonymous reports are stored in our 'Dreamcatcher' database.

Dream:ON was launched at the 2012 Edinburgh International Science Festival. Within a few hours I knew that we had a hit on our hands. The story quickly made its way around the globe, and was reported by the BBC, NBC, and CNN. Within days half a million people had downloaded the app and over time we have amassed millions of dream reports.

Lots of people use the app on a daily basis and so we are able to follow an ongoing storyline across several dreams. One of our most regular dreamers, for example, is having an imaginary affair with George Clooney. According to her dream reports, the two of them first met when George served her in a chemist shop, and asked if she would like to go for a walk with him and his giant giraffe.

Are people's dreams influenced by them choosing different

soundscapes? Absolutely. If someone chooses the nature land-scape they tend to experience dreams that involve greenery, flowers, and meadows. In contrast, when they select the beach soundscape they are far more likely to be transported to the coast, and suddenly feel the sun beating down on their skin. The effect may be due to the subtle sounds being played during the night, the power of suggestion, or a combination of the two. Whatever the explanation, we know that it is possible to help people shape their dreams and, encouraged by these findings, we have now created soundscapes based on *Fifty Shades of Grey* (don't ask), and another that contains fake news reports of a zombie apocalypse.

But does dream control always have to involve trying to become lucid and playing secret sounds in the middle of the night? Actually, no. In fact, millions of people have learnt how to use a surprisingly easy and effective way of shaping their noctur-nal fantasies. To find out more, we begin by discovering what not to think.

The dark side of the moon

People have long associated a full moon with all manner of strange behaviours. Indeed, the term 'lunatic' is derived from the Latin word for moon and has its roots in an old belief that people behaved in a bizarre way if the moon shone on them while they were asleep.

In 2013, neuroscientist Christian Cajochen from the University of Basel decided to take a scientific look at this rather strange idea.[12] Cajochen re-analysed the data from several experiments that had been conducted in a university sleep laboratory. During these studies, volunteers had been connected to an EEG machine and had their sleep monitored throughout the night. Cajochen looked at the date of each session, figured out where it fell in the lunar cycle, and then plotted this against their sleeping patterns. An interesting relationship quickly emerged.

Around the time of a full moon, the volunteers slept about twenty minutes less, took approximately five minutes longer to fall asleep, and experienced 30 per cent less deep sleep throughout the night.

What might explain this unusual effect? The patterns were not related to the menstrual cycle of female volunteers, and none of the participants could see the moon during the night. Instead, Cajochen favoured an evolutionary hypothesis, speculating that lighter sleeping patterns around the time of a full Moon might have helped our predecessors guard against predators who were able to see more clearly in the moonlight.

Fascinated by these findings, I examined whether the lunar cycle might also influence dreams. I randomly selected a few hundred dreams from the Dream:ON database and had them rated for bizarreness between '1' (such as a man walking into a bar and having a quiet drink) and '7' (for example, a horse walking into a bar and suddenly turning into a hot air balloon). I then plotted these ratings against the lunar cycle and was surprised to find that people were indeed having more bizarre dreams around the time of a full moon. So, if you want to enjoy especially strange dreams, make sure that you get an early night whenever there is a full moon.

On the rebound

Right now, millions of people across the world will be attempting to control their dreams. When people feel especially anxious or worried, they often have bad dreams or even nightmares. In an attempt to prevent these unpleasant experiences, people often try to push these concerns out of their mind as they fall asleep. In doing so, they may well be making the situation far worse.

Harvard psychologist Daniel Wegner has spent much of his career investigating the so-called 'rebound effect', wherein people who are asked not to think about something simply can't keep it out of their mind.[13] In my previous book, *Paranormality*, I described how his best-known experiment involved asking volunteers *not* to think about a white bear, and then getting them to indicate each time the aforementioned bear sauntered into their mind. The volunteers struggled to keep their brains bare of bears, and instead reported that the unwelcome creature kept popping up time and again. Why does this curious phenomenon occur? According to some psychologists, when you try to push a thought out of your mind, you often end up repeatedly asking yourself 'Am I thinking about what I am not supposed to be thinking about?' Because of this, you constantly remind yourself about the very thing that you are trying to forget, making it tricky to banish the thought from your mind.

The rebound effect affects many areas of our lives, including when we head for bed. In one study, for example, insomniacs were asked to try to forget whatever it was that was keeping them awake. As a result, they had an even harder time than usual falling asleep.[14] The effect can also exert an equally powerful influence over our dreams.

In 2011, psychologist Richard Bryant, from the University of New South Wales, asked volunteers to briefly think about an unpleasant memory just before they went to sleep.[15] Half of the volunteers were then asked to spend five minutes jotting down the thoughts and ideas that went through their head, but to actively avoid thinking about their unpleasant memory. In contrast, the other volunteers spent the same amount of timing jotting down their thoughts, and were allowed to think about their unpleasant memory. The following morning, all of the volunteers reported the dreams that they had had during the night, and the researchers counted the numbers of times that each volunteer mentioned their unpleasant memory. The results revealed that asking the volunteers to try *not* to think about the unpleasant memory prior to going to sleep doubled the likelihood of the unwanted memory popping up in their dreams.

Understanding the rebound effect can help those wanting to prevent bad dreams and nightmares. Rather than trying not to think about your worries and concerns before you go to sleep, spend a few minutes allowing these thoughts to freely flow through your mind. Don't actively encourage them, but instead just let them come in one ear and out the other. This technique will help make your dreams more pleasant. However, when it comes to taking control of the more negative side of the night, there is an even easier way to make nightmares vanish into thin air.

How to banish nightmares

Around 90 per cent of the population has at least one nightmare each month, and around 3 per cent of children and 1 per cent of adults report that they suffer from recurring nightmares. These terrifying experiences can cause people to suffer from a variety of night-time problems, including insomnia, fragmented sleep, and even sleep apnoea. However, the good news is that help is at hand.

Sleep expert Barry Krakow has dedicated much of his career to helping people cope with these highly emotional experiences, and has developed a three-stage technique known as 'imagery rehearsal therapy' (see page 285). Krakow argues that many bad dreams and nightmares are a form of nocturnal habit, and that it is possible to re-train people's brains to view the episode in a different way.

In the first stage of imagery rehearsal therapy, people are asked to select an unpleasant dream or nightmare to work with. Second, the person is encouraged to think of ways in which they would like the dream to change. This might, for example, involve altering the way in which they behave in the dream, or the ending of the episode. Finally, they are asked to rehearse the new version of the dream in their mind.

Krakow has discovered that some people struggle with the technique because they believe that their bad dreams are coming from their unconscious mind, and that it is not possible to change the way in which their unconscious operates. This 'once a smoker, always a smoker' approach to nightmares is also wrong. As we have seen time and again throughout *Night School*, both the quality and quantity of sleep and dreaming can be radically improved by making the smallest of changes. Krakow and his team have shown that imagery rehearsal therapy is highly effective, as long as people are able to overcome this concern and use the technique for a few weeks. The vast majority of people can banish the bogeyman from their bedroom, with studies showing that the technique has a remarkable 90 per cent success rate.[16] Imagery rehearsal therapy can also help reduce many of the symptoms associated with post-traumatic stress disorder, as well as helping to quell the unpleasant dreams often associated with living through a natural disaster[17] or being sexually assaulted.[18]

Imagery rehearsal therapy

This technique involves the following three steps.*

1) Determine the storyline of your nightmare

You first need to figure out the narrative underlying your nightmare. If you are experiencing several nightmares, you might want to start by choosing one that doesn't make you feel too anxious. Next, write down a first-person, present-tense account of the experience ('First, I find myself trapped in a cave . . . , etc). When you start to do this, you will probably focus on the scariest elements, such as the monster that attacks you or the threatening man that is always following you. That's fine. However, try to map out as much of the storyline as possible. Think about where you are in the nightmare, the time of day, and who is with you. Build up as many details as possible. You might find this somewhat scary, but remember to take it step by step, and stop if you feel overly anxious.

---

* This exercise is designed to illustrate a technique used by health professionals. If you believe that you, or your child, have a psychological problem, please consult a professional.

2) Rewrite the storyline

This stage is all about re-writing the nightmare. Again, using the first person and present tense create a much more pleasant version of events. Perhaps it turns out that the scary events are just part of a film that you are starring in, and stop the moment that the director shouts 'cut'. Or perhaps the scary people in the nightmare turn out to be your friends. Or maybe you don't run down that dark alley, but instead head for a well-lit street. Try to keep the scenario positive and believable.

3) Rehearse it

Now it is time for some imagery-based rehearsal. Find a quiet place to lie down during the day. Close your eyes, relax, and try to imagine your new storyline in as much detail as possible. Feel free to invent new and fun details. The important thing to remember is that you shouldn't be thinking about what is happening, but actually trying to form the scene in your head. Try to make the images and sounds as vivid as those in your dreams. If an unwanted element pops into your head, open your eyes, take a deep breath, recognize the intrusion, and then go back into the exercise. Try to carry out the exercise at least twice a day, for about three minutes each time.

Finally, if you have the nightmare again, your new and improved version of events will be much more likely to pop into your mind, and so transform the negative experience into something far more pleasant.

Psychotherapist Ann Sayre Wiseman (what a lovely surname) has examined how the same type of approach can be used to help children who are troubled by bad dreams and nightmares.[19] As with imagery rehearsal therapy, Wiseman's technique involves several simple stages, with children being encouraged to draw and talk about their bad dream, think about how they could transform it into something more pleasant, and then mentally rehearse this new and improved dream (see below).

Children's nightmares

Many children experience nightmares, and these bad dreams usually involve them being attacked by some kind of monster or negative entity. If your child repeatedly experiences the same nightmare, try the following:[*]

- When they wake up from a distressing nightmare, give them a hug and offer some reassurance. Rather than discuss the experience then and there, say something to lower their expectations of the experience happening again that night (perhaps monsters need to sleep and so only tend to appear once a night), and that you will talk about the dream tomorrow.

- The next day, encourage your child to make some kind of artwork that helps to illustrate the nightmare. This could be a drawing or a model. Ask them to use this artwork as a starting point to describe their bad dream. If they are scared about discussing what happened, you could suggest that they add some sort of self-protection to their

---

[*] This exercise is designed to illustrate a technique used by health professionals. If you believe that you, or your child, have a psychological problem, please consult a professional.

drawing or model, such as a shield, a cage around the monster, an army of friends, or a superhero.

- Ask your child how they might change the bad dream into something far more positive. Many children might suggest killing the monster. This isn't a great idea because it encourages them to dream about violence, and suggests that physical force is the best way of solving a problem. Instead, encourage them to humanize the monster, to understand what it wants, and to see if it has something important to say.

- Finally, encourage your child to draw, or model, a far more positive version of their dream. It is especially important to have them consider the kindest, and wisest, way of responding to the monster in the dream. This might involve them imagining a much more friendly relationship with the monster, drawing on the help of a superhero, a wise and strong animal, a parent or a teacher. Encourage your child to rehearse this new scenario a few times, and suggest that this version of events is likely to simply pop into their head if the nightmare happens again.

A dream life

A large body of work shows that people can use the power of imagination to change their dreams. I wondered whether it might be possible to take things one stage further and use this idea to help them improve their waking lives. It was time to carry out the first ever experiment to explore dreaming and self-development.

I assembled a group of 400 volunteers who wanted to achieve

an important aim, such as losing weight, eating more healthily, stopping smoking, or improving their career. I tracked the volunteers for two weeks, and then asked everyone whether they had experienced at least one goal-related dream and if they had made significant progress towards their aim. My theory was that the volunteers who had dreamt about their goal would be more successful than the others. The results were fascinating. Overall, it was a high-achieving group, with 58 per cent of volunteers making significant progress towards their goal. As predicted, dreaming played a key role in the process, with 74 per cent of those dreaming about their goals making significant strides towards achieving them, compared to only 40 per cent of those that didn't have such dreams. I had found the first scientific evidence that dreaming plays a key role in helping people improve their waking lives.

The experiment also identified the types of people who are likely to have goal-related dreams. Before the study started, everyone was asked to rate their imagery skills by indicating how easy they found it to create a mental picture of an object or scene. On the basis of these ratings, everyone was categorized as either a 'high' or 'low' visualizer. We found that 32 per cent of the high visualizers dreamt about their goal, versus only 12 per cent of the low visualizers. In short, proof that some people are born dreamers and further evidence that dreams can help people achieve their aims.

The final part of the study also started to explore whether it might be possible to encourage people to dream about their goals and achieve them. All of the volunteers were randomly assigned to one of three groups. One group acted as control and didn't receive any special advice about how to achieve their goals. The second group was asked to spend a few moments in the middle of the day visualizing themselves behaving in a way that was consistent with their aim. For example, if a volunteer wanted to eat more healthily, they might be asked to imagine themselves choosing a piece of fruit instead of a slice of cake. Similarly, if they wanted to stop smoking, they might imagine themselves using an electronic cigarette whenever they felt the urge to light up. The final group was asked to complete the same visualization exercise, but instead of forming the mental images towards the middle of the day they were asked to try to influence their dreams by visualizing just before they went to sleep. The results again showed the power of dream visualization. Compared to all of the other volunteers, those in the 'visualize before bedtime' group who did indeed shape their dreams were 10 per cent more likely to actually achieve their goals.

The study shows that people really can make their dreams a reality. This radically new approach to change has important implications for every area of our personal and professional lives. Although important, it is only the start of the journey. Future work needs to explore how everyone can best incorporate their aims and ambitions into their dreams. From sport to self-help, entrepreneurship to education, it is time to harness the power of the night.

For centuries people struggled to control their dreams. Now, the science of sleep can help people shape their nocturnal fantasies. As a result, millions of people across the world are able to create their perfect dream, transform their nightmares into sweet dreams, and change their lives while they are sound asleep. Dream control has moved from the realms of science fiction into science fact.

Conclusion

TIME FOR BED

Where we engage in some myth-busting, reveal ten things
that every adult and child in the country should know about
sleep and dreaming, and set out to change the world.

At the start of my talks about Night School, I regularly present audiences with each of the following statements, and ask them to raise their hands if they think the statement is true.

1) When I am asleep, my brain switches off.

2) I can learn to function well on less sleep.

3) Napping is a sign of laziness.

4) Snoring is annoying, but harmless.

5) I know when I am getting sleepy.

6) Dreams consist of meaningless thoughts and images.

7) Sleep is for wimps, and productive people spend less time in bed.

8) I don't have a problem getting enough sleep, and so there's no need to try to improve things.

9) A small amount of alcohol before bedtime improves sleep.

10) I can catch up on my lost sleep at the weekend.

11) Teenagers who spend lots of time in bed are just being lazy.

12) Eating cheese just before you go to bed gives you nightmares.

All of the statements suggest that sleep and dreaming are of little value, and that it's possible to treat tiredness with pills and the odd lie in. During my talks, each of them is greeted with a sea of hands. We have examined all of these topics in detail during Night School, so let's refresh our memories by taking a quick look at each of the statements in turn.

1) When I am asleep, my brain switches off

MYTH. When you fall asleep, your sense of self-awareness shuts down, and so it may feel as if you have become inactive. In fact, your brain remains highly active during sleep, and carries out several tasks that are essential for your well-being.

2) I can learn to function well on less sleep

MYTH. Sleep is a biological need, and it simply isn't possible to cut corners. Of course, you can force yourself to sleep less, but you will not be fully rested, and the way in which you think, feel, and behave will be impaired.

3) Napping is a sign of laziness

MYTH. Your circadian rhythm makes you feel sleepy throughout the night and for a much shorter period of time towards the middle of the afternoon. Napping is entirely natural, and helps make you more alert, focused, creative, and productive.

4) Snoring is annoying, but harmless

MYTH. Snoring can be a symptom of sleep apnoea. This serious medical condition causes you to experience hundreds of mini-awakenings throughout the night, and is associated with an increased risk of heart disease, obesity, and cancer.

5) I know when I am getting sleepy

MYTH. People are very poor judges of how tired they are. As a result, they often drive when they are drowsy, and struggle through the day not realizing that they are far from their best.

6) Dreams consist of meaningless thoughts and images

MYTH. When you dream, your brain is often attempting to work through your concerns. As a result, your dreams can provide a useful insight into your worries, and also help come up with innovative solutions to these issues.

7) Sleep is for wimps, and productive people spend less time in bed

MYTH. If you don't get enough sleep then you'll struggle to concentrate, become accident-prone, lack willpower, and become less productive. Worse still, you will increase your chances of becoming overweight, having a heart attack, and dying early.

8) I don't have a problem getting enough sleep, and so there's no need to try to improve things

MYTH. A small percentage of people are super-sleepers. They enjoy a good night's sleep almost every night, can fall asleep whenever they want, and have sweet dreams. Compared to most, they are happier, healthier, and wealthier. Even if your sleep is OK, you can still improve by becoming a super-sleeper.

9) A small amount of alcohol before bedtime improves sleep

MYTH. Alcohol may help you to fall asleep, but it also results in a far more disturbed night. Even a small tipple results in you spending less time in restorative deep sleep, having fewer dreams, and being more likely to snore.

10) I can catch up on my lost sleep at the weekend

MYTH. When you fail to get enough sleep you develop a sleep debt. Spending more time in bed for a day will help ease this problem, but won't fully restore you for the coming week. Over time this lifestyle results in many of the problems associated with sleep deprivation.

11) Teenagers who spend lots of time in bed are just being lazy
MYTH. When we hit adolescence our circadian rhythms become delayed by about three hours, causing us to become more 'evening types'. In addition, teenagers require between nine and ten hours' sleep each night. They are not being lazy, it is just biology at work.

12) Eating cheese just before you go to bed gives you nightmares
MYTH. This myth probably has its origins in Charles Dickens's story *A Christmas Carol*. In 2005, the British Cheese Board asked 200 volunteers to spend a week eating a small piece of cheese before going to sleep, and then report their dreams the next morning. None of them had nightmares.

All of the statements are incorrect. Nevertheless, the vast majority of the public are convinced that they are true and, as a result, place very little value on sleep and dreaming. This has had a terrible impact on society. Rather than celebrate the night, sleep and dreaming are now treated as annoying interruptions to our all-important lives. Living in a world that hasn't had a good night's rest for years has finally taken its toll. The vast majority of school children and students now arrive for their classes severely sleep deprived, adult sleep debt is at a record high, the demand for sleeping pills is rising year on year, and millions of people go about their daily business in a zombie-like state that is ruining their relationships, health and productivity.

Perhaps more than at any other point in history, there is now an urgent need to change our attitudes towards the night. I believe that this will require nothing short of a revolution.

During *Night School* we have examined the science surrounding sleep and dreaming. In the first few lessons we discovered what your brain gets up to every night of your life, the severe downside of sleep deprivation, the secrets of super-sleep, the best way of banishing the bogeyman from your bedroom, and the surprising truth about sleep learning and power napping. In latter lessons, we found out what really happens in your dreams, discovered how dreams can provide a useful insight into your worries and concerns, and learnt how to experience your perfect nocturnal fantasy. Unfortunately, up until now, much of this material has remained hidden away in academic journals. This needs to change. In the same way that we want every adult and child in the country to have key literacy and numeracy skills, so we also need to ensure that they understand the fundamental findings from sleep science. Although extremely important, this is only part of the solution.

Throughout *Night School* we have encountered a series of quick, but highly effective, techniques that help people get a great night's sleep and learn from their dreams. These techniques can help people move from poor sleepers to good, and from good to great. It's now vital that these techniques become common knowledge. To this end, I have assembled 'The Night School Manifesto' – a list of ten of the most effective Night School techniques (see overleaf). My hope is that over the next few years this manifesto will become part of the national curriculum, and regularly feature on the media and the Internet. If this dream becomes a reality, millions of people will be able to improve their lives by making the most of the night.

The Night School Manifesto

Here are ten techniques that will allow everyone to get the most out of the night.*

If you want to feel sleepy when you head to bed . . .
Banish the blues: Try wearing glasses with amber-tinted lenses for two to three hours before you go the bed. These will block the blue light that stimulates your brain and make you feel especially sleepy.

If you want to feel especially refreshed in the morning . . .
The 'ninety-minute' rule: Decide when you want to wake up, and then count back in ninety-minute blocks (the length of a sleep cycle) to discover the best to time to head to bed. For example, if you want to wake up at 8 a.m., go to bed around 11 p.m. or 12.30 a.m.

If you want to fall asleep quickly . . .
Use positive imagery and the paradox principle: First, imagine yourself in a very pleasant scenario. Make the scenario as detailed as possible, but avoid anything that's too exciting, perhaps planning your dream holiday or thinking about your perfect evening out. If that doesn't work, try to stay awake. Strange as it sounds, forcing yourself to actually remain awake is one of the best ways of nodding off.

---

* This exercise is designed to illustrate techniques used by health professionals. If you believe that you, or your child, have a psychological problem, please consult a professional.

If you lie in bed feeling worried . . .

The list: If you have a lot on your mind, make a list of all of the things that you have to do the next day. If you are worrying about something specific, jot that down too and try to allow the thought to drift through your mind, rather than focusing on it.

If you wake up in the middle of the night . . .

The jigsaw method: You might be experiencing a perfectly natural phenomenon known as 'segmented sleep', where people sleep in two long blocks, with a gap of roughly thirty minutes between them. However, if you lie awake for more than twenty minutes, get up and do something non-stimulating for a few minutes, such as working on a jigsaw.

If you want to learn in your sleep . . .

The real secret of sleep learning: Sleep glues memories into your mind, so don't stay up late trying to cram information into your brain. Instead, study during the day, remind yourself about key points just before you nod off, and get lots of sleep at night.

If you want to boost your brainpower during the day . . .

Neuro-napping: Taking a catnap will help you to become more alert, creative, and productive. Neuro-napping involves listening to music when you are studying or brainstorming, and then playing the same music when you nap. Napping boosts memory and creativity by around 60 per cent.

If you are experiencing a recurring nightmare, or bad dream . . .

Imagery rehearsal therapy: Spend some time during the day describing your nightmare, creating a different ending for the episode, and then imagining this new and improved ending. Studies show that this simple technique stops nightmares 90 per cent of the time.

If you want to gain an insight into your concerns and worries . . .
Dream work: Describe a striking dream in detail, look for ways in which it applies to your life, and then use this as the basis for change. Research shows that around 80 per cent of people find that this yields an important insight into their concerns.

If you want to achieve a goal . . .
The power of pre-sleep suggestion: Just before you fall asleep, imagine doing whatever you need to do to achieve your goal. For example, if you want to go to the gym more frequently, imagine yourself putting on your trainers and heading out the house. As you drift asleep, tell yourself that you want these images to crop up in your dreams.

But even this is still just the tip of the iceberg. Almost every aspect of society needs to undergo a radical change in focus. Clinicians and medics should appreciate how the new science of sleep can be used to promote physical health and overcome psychological disorders. Businesses across the globe need to realize how they can boost productivity by encouraging employees to take a short sleep in cost-effective nap rooms. Hotels and airports could create effective ways of helping weary travellers overcome the devastating effects of jet lag. Schools and colleges need to ensure that every child understands how to get a good night's rest, learn while they are sound asleep, and eliminate nightmares. In short, night needs to become the new day.

For years, the self-development movement has focused on improving people's waking lives. It's time to celebrate the new science of sleep and make the most of the missing one-third of every day. It's time to embrace the night. Perhaps most important of all, it's time for bed. Sleep well.

Notes

Introduction: Waking up

1 This questionnaire is based on work presented in: Horne, J., & Östberg, O. (1976). 'A Self-Assessment Questionnaire to Determine Morningness-Eveningness in Human Circadian Rhythms'. *International Journal of Chronobiology*, 4, 97–110.

Lesson 1: Into the night

1 For further information about Berger, see: Millett, D. (2001). 'Hans Berger: From Psychic Energy to the EEG'. *Perspectives in Biology and Medicine*, 44(4), 522–42.

2 For more information about Loomis, see: Conant, J. (2002). *Tuxedo Park: A Wall Street Tycoon and the Secret Palace of Science that Changed the Course of World War II*. New York: Simon & Schuster.

3 Roberts, W. (2010). 'Facts and ideas from anywhere'. *Baylor University Medical Center Proceedings*, 23(3), 318–32.

4 For further information about Eugene Aserinsky, see: Aserinsky, E. (1996). 'The Discovery of REM Sleep'. *Journal of the History of the Neurosciences*, 5, 213–27.

5 Aserinsky, E., & Kleitman, N. (1953). 'Regularly Occurring Periods of Eye Motility, and Concomitant Phenomena, During Sleep'. *Science*, 118(3062), 273–4.

6 Murzyn, E. (2008). 'Do We Only Dream in Colour? A Comparison of Reported Dream Colour in Younger and Older Adults with Different Experiences of Black and White Media'. *Consciousness and Cognition*, 17(4), 1228–37.

NOTES

7 Fisher, C. (1966). 'Dreaming and Sexuality'. In *Psychoanalysis – A General Psychology*. New York: International Universities Press. 537–69.

8 Hurovitz, C., Dunn, S., Domhoff, G. W., & Fiss, H. (1999). 'The Dreams of Blind Men and Women: A Replication and Extension of Previous Findings'. *Dreaming*, 9, 183–93.

9 Rechtschaffen, A., & Foulkes, D. (1965). 'Effect of Visual Stimuli on Dream Content'. *Perceptual and Motor Skills*, 20, 1149–60.

10 To find out more about Jean-Jacques d'Ortous de Mairan, see: Roenneberg, T. (2012). *Internal Time: Chronotypes, Social Jet Lag, and Why You're So Tired*. Massachusetts: Harvard University Press.

11 De Mairan, J. J. O. (1729). 'Observation Botanique'. *Histoire de l'Academie Royale des Sciences*, 35.

12 Siffre, M. (1964). *Beyond Time*. New York: McGraw-Hill.

13 Stolarski, M., Ledzińska, M., & Matthews. G. (2013). 'Morning is Tomorrow, Evening is Today: Relationships Between Chronotype and Time Perspective'. *Biological Rhythm Research*, 44, 181–96.

14 Giampietro, M., & Cavallera, G. M. (2006). 'Morning and Evening Types and Creative Thinking'. *Personality and Individual Differences*, 42(3), 453–46.

15 Jonason, P. K., Jones, A., & Lyons, M. (2013). 'Creatures of the Night: Chronotypes and the Dark Triad Traits'. *Personality and Individual Differences*, 55(5), 538–41.

16 Preckel, F., Lipnevich, A., Boehme, K., Brandner, L., Georgi, K., Könen, T., Mursin, K., & Roberts, R. (2013). 'Morningness-Eveningness and Educational Outcomes: The Lark has an Advantage Over the Owl at High School'. *British Journal of Educational Psychology*, 83(1), 114–34.

17 Roenneberg, T. (2012). *Internal Time: Chronotypes, Social Jet Lag, and Why You're So Tired*. Massachusetts: Harvard University Press.

18 Recht, L. D., Lew, R. A., & Schwartz, W. J. (1995). 'Baseball Teams Beaten by Jet Lag'. *Nature*, 377, 583.

19 This questionnaire is based on various measures of sleep quality, including work described in: Buysse, D. J., Reynolds III, C. F., Monk, T. H., Berman, S. R., & Kupfer, D. J. (1989). 'The Pittsburgh Sleep Quality Index: A New Instrument for Psychiatric Practice and Research'. *Journal of Psychiatric Research*, 28(2), 193–213.

Lesson 2: How to be happy, healthy, wealthy, and wise

1 Martin, P. (2003). *Counting Sheep*. London: Flamingo.
2 Kripke, D., Simons, R., Garfinkel, L., et al. (1979). 'Short and Long Sleep and Sleeping Pills. Is Increased Mortality Associated?'. *Archives of General Psychiatry*, 36, 103–16.
3 National Sleep Foundation. *Sleep in America Poll, 2001–2002*. Washington, DC: National Sleep Foundation; and Spiegel, K., Leproult, R., & Van Cauter, E. (1999). 'Impact of Sleep Debt on Metabolic and Endocrine Function'. *Lancet*, 354, 1435–9.
4 Institute of Medicine (2006). *Sleep Disorders and Sleep Deprivation: An Unmet Public Health Problem*. Washington, DC: The National Academies Press.
5 Wheaton, A.G. (2011). 'Effect of Short Sleep Duration on Daily Activities: United States, 2005–2008'. *Morbidity and Mortality Weekly Report*, 60(08), 239–42.
6 Mental Health Foundation (2011). *Sleep Matters: The Impact Of Sleep On Health And Wellbeing*.
7 Rechtschaffen, A., Gilliland, M. A., Bergmann, B. M., & Winter, J. B. (1983). 'Physiological Correlates of Prolonged Sleep Deprivation in Rats'. *Science*, 221, 182–4.
8 For further information about Peter Tripp, see: Coren, S. (1996). *Sleep Thieves*. New York: Simon & Schuster.
9 Ross J. (1965). 'Neurological Findings After Prolonged Sleep Deprivation'. *Archives of Neurology*, 12, 399–403.
10 Dement, W. C. (1972). *Some Must Watch While Some Must Sleep: Exploring the World of Sleep*. New York: Norton.
11 http://opinionator.blogs.nytimes.com/2010/04/22/the-stay-awake-men/.
12 Channel 4 (2004). *Shattered*.
13 http://news.bbc.co.uk/1/hi/world/americas/8003537.stm.
14 Blagrove, M. (1996). 'Effects of Length of Sleep Deprivation on Interrogative Suggestibility'. *Journal of Experimental Psychology: Applied*, 2(1), 48–59.
15 Kamimori, G. H., Karyekar, C. S., Otterstetter, R., Cox, D. S., Balkin, T. J., Belenky, G. L., & Eddington, N. D. (2002). 'The Rate of

Absorption and Relative Bioavailability of Caffeine Administered in
Chewing Gum Versus Capsules to Normal Healthy Volunteers'.
International Journal of Pharmacology, 234, 159–67.

16 Belenky G., Wesensten N. J., Thorne D. R., et al. (2003). 'Patterns of
Performance Degradation and Restoration During Sleep Restriction
and Subsequent Recovery: A Sleep-Dose Response Study'. *Journal of
Sleep Research*, 12, 1–12.

17 http://www.nytimes.com/1990/02/25/us/teen-ager-honored-as-top-
us-driver-is-killed-in-accident.html.

18 http://abcnews.go.com/Technology/blink-eye-dozing-
driving/story?id=17870880.

19 Brain cells are fuelled by both glucose from the bloodstream and a
carbohydrate called glycogen. Although glycogen only provides a
small amount of this energy, it can quickly kick into action when the
brain suddenly becomes highly active (a bit like an emergency battery
during a power cut). Sleep deprivation prevents the build-up of
glycogen, and so robs the brain of the energy supply that it requires
to function properly. For more information, see: Benington, J. H., &
Heller, H. C. (1995). 'Restoration of Brain Energy Metabolism as the
Function of Sleep'. *Progress in Neurobiology*, 45, 347–60.

20 Harrison, Y., & Horne, J. A. (1998). 'Sleep Loss Affects Risk-Taking'.
Journal of Sleep Research, 7(2), 113.

21 Christian, M. S., & Ellis, A. P. J. (2011). 'Examining the Effects of
Sleep Deprivation on Workplace Deviance: A Self-Regulatory
Perspective'. *Academy of Management Journal*, 54(5), 913–34.

22 Barnes, C. M., Schaubroeck, J., Huth, M., & Ghumman, S. (2011).
'Lack of Sleep and Unethical Conduct'. *Organizational Behavior and
Human Decision Processes*, 115(2), 169–80.

23 Wagner, D. T., Barnes, C. M., Lim, V. K. G., & Ferris, D. L. (2012).
'Lost Sleep and Cyberloafing: Evidence from the Laboratory and a
Daylight Saving Time Quasi-Experiment'. *Journal of Applied
Psychology*, 97(5), 1068–76.

24 National Sleep Disorders Research Plan (2001). 'Sleep Disorders
Create Growing Opportunities for Hospitals'. *Health Care Strategy
Management*, 19(2), 16–17.

25 Ferrie, J. E., Shipley, M. J., Akbaraly, T. N., Marmot, M. G., Kivimäki,
M., & Singh-Manoux, A. (2011). 'Change in Sleep Duration and

Cognitive Function: Findings from the Whitehall II Study'. *Sleep,* 34(5), 565–73.

26 Gallicchio, L., & Kalesan, B. (2009). 'Sleep Duration and Mortality: A Systematic Review and Meta-Analysis'. *Journal of Sleep Research*, 18, 148–58.

27 Ferrie, J. E., Shipley, M. J., Cappuccio, F. P., Brunner, E., Miller, M. A., Kumari, M., & Marmot, M. G. (2007). 'A Prospective Study of Change in Sleep Duration: Associations with Mortality in the Whitehall II Cohort'. *Sleep*, 30(12) 1659–66.

28 Kripke, D. F., Garfinkel, L., Wingard, D. L., Klauber, M. R., & Marler, M. R. (2002). 'Mortality Associated with Sleep Duration and Insomnia'. *Archives of General Psychiatry*, 59, 131–6.

29 Brown, D., et al. (2009). 'Rotating Night Shift Work and the Risk of Ischemic Stroke'. *American Journal of Epidemiology*, 169, 1370–7.

30 Davis, S., Mirick, D., & Stevens, R. (2001). 'Night Shift Work, Light at Night, and Risk of Breast Cancer'. *Journal of the National Cancer Institute*, 93, 1557–62; and Schernhammer, E., Laden, F., Speizer, F., et al. (2001). 'Rotating Night Shifts and Risk of Breast Cancer in Women Participating in the Nurses' Health Study'. *Journal of the National Cancer Institute*, 93, 1563–8.

31 Schernhammer, E., et al. (2003). 'Night-shift Work and Risk of Colorectal Cancer in the Nurses' Health Study'. *Journal of the National Cancer Institute*, 95, 825–8.

32 Scheer, F., et al. (2009). 'Adverse Metabolic and Cardiovascular Consequences of Circadian Misalignment'. *Proceedings of the National Academy of Sciences*, 106, 4453–8; and Knutson, K., & Van Cauter, E. (2008). 'Associations Between Sleep Loss and Increased Risk of Obesity and Diabetes'. *Annals of the New York Academy of Sciences*, 1129, 287–304.

33 Luckhaupt, S. E (2012). 'Short Sleep Duration Among Workers – United States, 2010'. *Morbidity and Mortality Weekly Report*, 61(16), 281–285.

34 Ogden, C. L., Carroll, M. D., Kit, B. K., Flegal, K. M. (2012). 'Prevalence of Obesity in the United States, 2009–2010'. NCHS data brief, no 82. Hyattsville, MD: National Center for Health Statistics.

35 Reilly, J. J., Armstrong, J., Dorosty, A. R., et al. (2005). 'Early Life Risk

Factors for Obesity in Childhood: Cohort Study'. *British Medical Journal*, 330, 1357.

36 Patel, S. R., Malhotra, A., White, D. P., Gottlieb, D. J., & Hu, F. B. (2006). 'Association Between Reduced Sleep and Weight Gain in Women'. *American Journal of Epidemiology*, 164, 947–54.

37 Taheri, S., Lin L., Austin, D., Young, T., & Mignot, E. (2004). 'Short Sleep Duration is Associated with Reduced Leptin, Elevated Ghrelin, and Increased Body Mass Index'. *PLOS Medicine*, 1(3), e62, doi: 10.1371/journal.pmed.0010062.

38 Spiegel, K., Tasali, E., Penev, P., & Van Cauter, E. (2004). 'Brief Communication: Sleep Curtailment in Healthy Young Men is Associated with Decreased Leptin Levels, Elevated Ghrelin Levels, and Increased Hunger and Appetite'. *Annals of Internal Medicine*, 141, 846–50.

39 Chapman, C. D., Nilsson, E. K., Nilsson, V. C., et al. (2013). 'Acute Sleep Deprivation Increases Food Purchasing in Men'. *Obesity*, doi: 10.1002/oby.20579.

40 Axelsson, J., Sundelin, T., Ingre, M., et al. (2010). 'Beauty Sleep: Experimental Study on the Perceived Health and Attractiveness of Sleep Deprived People'. *British Medical Journal*, 341, 6614.

41 Sundelin, T., Lekander, M., Kecklund, G., Van Someren, E. J. W., Olsson, A., & Axelsson, J. (2013). 'Cues of Fatigue: Effects of Sleep Deprivation on Facial Appearance'. *Sleep,* 36(9), 1355–60.

42 Wulff, K., Dijk, D. J., Middleton, B., Foster, R. G., & Joyce, E. M. (2012). 'Sleep and Circadian Rhythm Disruption in Schizophrenia'. *British Journal of Psychiatry*, 200, 308–16.

43 Owens, J. A. (2005). 'The ADHD and Sleep Conundrum: A Review'. *Journal of Developmental and Behavioral Pediatrics*, 26, 312–22.

44 Xie, L., Kang, H., Xu, Q., Chen, M. J., Liao, Y., Thiyagarajan, M., O'Donnell, J., Christensen, D. J., Nicholson, C., Iliff, J. J., Takano, T., Deane, R., & Nedergaard, M. (2013). 'Sleep drives metabolite clearance from the adult brain'. *Science,* 342 (6156), 373–7.

45 Dunkell, S. (1977). *Sleep Positions: The Night Language of the Body.* London: Heinemann.

46 Domino, G., & Bohn, S. A. (1980). 'Hypnagogic Exploration: Sleep Positions and Personality'. *Journal of Clinical Psychology*, 36, 760–2; Schredl, M. (2002). 'Sleep Positions and Personality: An Empirical

Study'. *North American Journal of Psychology*, 4, 129–32; and Kamau, L. Z., Luber, E., & Kumar, V. K. (2012). 'Sleep Positions and Personality: Zuckerman–Kuhlman's Big Five, Creativity, Creativity Styles, and Hypnotizability'. *North American Journal of Psychology*, 14, 609–20.

Lesson 3: The secret of super-sleep

1 Stampi C., ed. (1992). *Why We Nap: Evolution, Chronobiology, and Functions of Polyphasic and Ultrashort Sleep*. Boston, MA: Birkhauser.
2 http://www.outsideonline.com/fitness/endurance-training/Miles-to-Go-Before-I-Sleep.html?page=all.
3 Meddis, R. (1977). *The Sleep Instinct*. Boston: Routledge & Kegan Paul.
4 Meddis, R., Pearson, A. J. D., & Langford, G. (1973). 'An Extreme Case of Healthy Insomnia'. *Electroencephalography and Clinical Neurophysiology*, 35, 213–14.
5 http://www.businessinsider.com/successful-people-who-barely-sleep-2012-9?op=1.
6 He, Y., Jones, C. R., Fujiki, N., Xu, Y., Guo, B., Holder, J. L., Rossner, M. J., Nishino, S., & Fu, Y. H. (2009). 'The Transcriptional Repressor DEC2 Regulates Sleep Length in Mammals'. *Science*, 325(5942), 866–70.
7 Horne, J. (2007). *Sleepfaring: A journey through the science of sleep*. Oxford: Oxford University Press.
8 McIntyre, I. M., Norman, T. R., Burrows, G. D., & Armstrong, S. M. (1989). 'Human Melatonin Suppression by Light is Intensity Dependent'. *Journal of Pineal Research*, 6, 149–56.
9 Sasseville, A., Paquet, N., Sévigny, J., & Hébert, M. (2006). 'Blue Blocker Glasses Impede the Capacity of Bright Light to Suppress Melatonin Production'. *Journal of Pineal Research*, 41, 73–8; and Burkhart, K., & Phelps, J. R. (2009). 'Amber Lenses to Block Blue Light and Improve Sleep: A Randomized Trial'. *Chronobiology International*, 26, 1602–12.
10 http://www.telegraph.co.uk/news/newstopics/howaboutthat/6684362/Buzzing-flies-more-likely-to-wake-men-than-crying-babies-study.html.
11 Miedema, H. M., & Vos, H. (2007). 'Associations Between Self-

Reported Sleep Disturbance and Environmental Noise Based on Reanalyses of Pooled Data From 24 Studies'. *Behavioral Sleep Medicine*, 5, 1–20.

12 Montgomery-Downs, H. E., Insana, S. P., & Miller, E. A. (2009). 'Effects of Two Types of Ambient Sound During Sleep'. *Behavioral Sleep Medicine*, 8, 40–7.

13 Buman, M. P., & King, A. C. (2010). 'Exercise as a Treatment to Enhance Sleep'. *American Journal Of Lifestyle Medicine*, 4, 500–14.

14 Horne, J. A., & Minard, A. (1985). 'Sleep and Sleepiness Following a Behaviourally Active Day'. *Ergonomics*, 28, 567–75.

15 Kyle, S. D., Morgan, K., Spiegelhalder, K., & Espie, C. A. (2011). 'No Pain, No Gain: An Exploratory Within-Subjects Mixed-Methods Evaluation of the Patient Experience of Sleep Restriction Therapy (SRT) for Insomnia'. *Sleep Medicine*, 12, 735–747.

16 Carney, C. E., & Waters, W. F. (2006). 'Effects of a Structured Problem-Solving Procedure on Pre-Sleep Cognitive Arousal in College Students with Insomnia'. *Behavioral Sleep Medicine*, 4(1), 13–28.

17 Stone, B. (1980). 'Sleep and Low Doses of Alcohol'. *Electroencephalography and Clinical Neurophysiology*, 48, 706–9.

18 http://latimesblogs.latimes.com/nationnow/2011/11/thanksgiving-busting-the-tryptophan-myth-wide-open.html.

19 Hirokawa, K., Nishimoto, T., & Taniguchi, T. (2012). 'Effects of Lavender Aroma on Sleep Quality in Healthy Japanese Students'. *Perceptual and Motor Skills*, 114, 111–22; and Lewith, G. T., Godfrey, A. D., & Prescott, P. (2005). 'A Single-Blinded, Randomized Pilot Study Evaluating the Aroma of Lavandula Augustifolia as a Treatment for Mild Insomnia'. *Journal of Alternative and Complementary Medicine*, 11, 631–7.

20 Haynes, S. N., Adams, A., & Franzen, M. (1981). 'The Effects of Presleep Stress on Sleep-Onset Insomnia'. *Journal of Abnormal Psychology*, 90, 601–6.

21 Harvey, A. G., & Payne, S. (2002). 'The Management of Unwanted Pre-Sleep Thoughts in Insomnia: Distraction with Imagery Versus General Distraction'. *Behaviour Research and Therapy*, 40, 267–77.

22 Broomfield, N. M., & Espie, C. A. (2003). 'Initial Insomnia and Paradoxical Intention: An Experimental Investigation of Putative

Mechanisms Using Subjective and Actigraphic Measurement of Sleep'. *Behavioural and Cognitive Psychotherapy*, 31, 313–24.

23 Tang, N. K. Y., & Harvey, A. G. (2004). 'Correcting Distorted Perception of Sleep in Insomnia: A Novel Behavioural Experiment?'. *Behaviour Research and Therapy*, 42, 27–39.

24 Mercer, J. D., Bootzin, R. R., & Lack, L. C. (2002). 'Insomniacs' Perception of Wake Instead of Sleep'. *Sleep*, 25, 564–71.

25 Tang, N. K. Y., & Harvey, A. G. (2004). 'Correcting Distorted Perception of Sleep in Insomnia: A Novel Behavioural Experiment?'. *Behaviour Research and Therapy*, 42, 27–39.

26 Ekirch, A. R. (2005). *At Day's Close: Night in Times Past*. New York: W.W. Norton.

27 Wehr, T. A. (1992). 'In Short Photoperiods, Human Sleep is Biphasic'. *Journal of Sleep Research*, 1, 103–7.

28 Mindell, J. A., Telofski, L. S., Wiegand, B., & Kurtz, E. S. (2009). 'A Nightly Bedtime Routine: Impact on Sleep in Young Children and Maternal Mood'. *Sleep*, 32, 599–606.

29 Rickert, V. I., & Johnson, C. M. (1988). 'Reducing Nocturnal Awakening and Crying Episodes in Infants and Young Children: A Comparison Between Scheduled Awakenings and Systematic Ignoring'. *Pediatrics*, 81, 203–12.

30 Adams, L. A., & Rickert, V. I. (1989). 'Reducing Bedtime Tantrums: Comparison Between Positive Routines and Graduated Extinction'. *Pediatrics*, 84, 756–61.

31 Carpenter, R., McGarvey, C., Mitchell, E. A., et al. (2013). 'Bed Sharing when Parents do not Smoke: Is there a Risk of SIDS? An individual level analysis of five major case–control studies'. *BMJ* Open, doi:10.1136/bmjopen-2012-002299.

Lesson 4: On sleepwalking and night terrors

1 http://www.thesun.co.uk/sol/homepage/news/104719/Its-a-real-Flash-Garden.html.
2 http://news.bbc.co.uk/1/hi/uk/4654579.stm.
3 http://www.dailymail.co.uk/news/article-381584/Ex-chef-cooks-meals-sleep.html.

4 http://www.newscientist.com/article/dn6540.

5 http://www.dailymail.co.uk/news/article-528574/Meet-Kipasso---
sleepwalking-nurse-draws-masterpieces-trance.html.

6 http://www.dailymail.co.uk/news/article-2262188/Its-Vincent-van-
Sloth-Artist-Lee-Hadwin-draw-awake-creates-masterpieces-SLEEP-
unveils-1m-collection.html.

7 Siddiqui, F., Osuna, E., & Chokroverty, S. (2009). 'Writing emails as
part of sleepwalking after increase in Zolpidem'. *Sleep Medicine*, 10(2),
262–4.

8 Ohayon, M. M., Mahowald, M. W., Dauvilliers, Y., Krystal, A. D., &
Léger, D. (2012). 'Prevalence and Comorbidity of Nocturnal
Wandering in the US Adult General Population'. *Neurology*, 78,
1583–9.

9 http://knarf.english.upenn.edu/Articles/rieger.html#25.

10 See, for example, Licis, A. K., Desruisseau, D. M., Yamada, K. A.,
Duntley, S. P., & Gurnett, C. A. (2011). 'Novel Genetic Findings in an
Extended Family Pedigree with Sleepwalking'. *Neurology*, 76, 49–52.

11 Guilleminault, C., Moscovitch, A., & Leger, D. (1995). 'Forensic Sleep
Medicine: Nocturnal Wandering and Violence'. *Sleep*, 18, 740–8.

12 Hufford, D. J. (1982). *The Terror That Comes in the Night*. Philadelphia:
University of Pennsylvania Press; and Kotorii, T., Kotorii, T.,
Uchimura, N., Hashizume, Y., Shirakawa, S., Satomura, T., et al.
(2001). 'Questionnaire Relating to Sleep Paralysis'. *Psychiatry and
Clinical Neurosciences*, 55, 265–6.

13 Bassetti, C., Vella, S., Donati, F., Wielepp, P., & Weder, B. (2000),
'SPECT During Sleepwalking'. *Lancet*, 356(9228), 484–5.

14 Pilon, M., Montplaisir, J., & Zadra, A. (2008). 'Precipitating Factors of
Somnambulism: Impact of Sleep Deprivation and Forced Arousals'.
Neurology, 70, 2284–90.

15 Pressman, M. R. (2009). 'Sleepwalking Déjà Vu' (commentary on
Oudiette, et al. 'Dreamlike mentations during sleepwalking and sleep
terror in adults'). *Sleep*, 32(12), 1621–7.

16 Ohayon, M. M., Mahowald, M. W., Dauvilliers, Y., Krystal, A. D., &
Léger, D. (2012). 'Prevalence and Comorbidity of Nocturnal
Wandering in the US Adult General Population'. *Neurology*, 78,
1583–9; and Kales, J. D., Kales, A., Soldatos, C. R., Caldwell, A.,
Charney, D., & Martin, E. D. (1980). 'Night Terrors: Clinical

Characteristics and Personality Patterns'. *Archives of General Psychiatry*, 37, 1413–17.

17 Guilleminault, C., Moscovitch, A., & Leger, D. (1995). 'Forensic Sleep Medicine: Nocturnal Wandering and Violence'. *Sleep*, 18, 740–8.

18 Pressman, M. R. (2007). 'Disorders of Arousal from Sleep and Violent Behavior: The Role of Physical Contact and Proximity'. *Sleep*, 30, 1039–47.

19 Ficker, J., Wiest, G., Lehnert, G., Meyer, M., & Hahn, E. (1999). 'Are Snoring Medical Students at Risk of Failing their Exams?'. *Sleep*, 22, 205–9.

20 Terán-Santos, J., Jimenez-Gomez, A., & Cordero-Guevara, J. (1999). 'The Association Between Sleep Apnea and the Risk of Traffic Accidents'. *New England Journal of Medicine*, 340, 847–51; and Findley, L., Unverzagt, M., Guchu, R., Fabrizio, M., Buckner, J., & Suratt, P. (1995). 'Vigilance and Automobile Accidents in Patients with Sleep Apnea or Narcolepsy'. *Chest*, 108, 619.

21 Guimarães, K. C., Drager, L. F., Genta, P. R., Marcondes, B. F., & Lorenzi-Filho, G. (2009). 'Effects of Oropharyngeal Exercises on Patients with Moderate Obstructive Sleep Apnea Syndrome'. *American Journal of Respiratory and Critical Care Medicine*, 179, 962–6.

22 Puhan, M. A., Suarez, A., Lo Cascio, C., Zahn, A., Heitz, M., & Braendli, O. (2006). 'Didgeridoo Playing As Alternative Treatment For Obstructive Sleep Apnoea Syndrome: Randomised Controlled Trial'. *British Medical Journal*, 332(7536), 266–70.

23 Ojay, A., & Ernst, E. (2000). 'Can Singing Exercises Reduce Snoring? A Pilot Study'. *Complementary Therapy Medicine*, 8, 151–6.

Lesson 5: Sleep learning and power naps

1 Leshan, L. (1942). 'The Breaking of a Habit by Suggestion During Sleep'. *Journal of Abnormal Social Psychology*, 37, 406–8.

2 Rubin, F., ed. (1964). *Current Research in Hypnopaedia*. New York: American Elsevier, 77.

3 Bliznitchemko, L. (1968). 'Hypnopaedia and its practice in the USSR'. pp. 202–209 in Rubin, F., ed. (1964). *Current Research in Hypnopaedia*. New York: American Elsevier.

4 Emmons, W. H., & Simon, C. W. (1956). 'The Non-Recall of Material Presented During Sleep'. *American Journal of Psychology, 69*, 76–81.

5 Ebbinghaus, H. (1885). *Memory: A Contribution to Experimental Psychology.* New York: Teachers College, Columbia University.

6 See, for example, Jenkins, J. G., & Dallenbach, K. M. (1924). 'Obliviscence During Sleep and Waking'. *The American Journal of Psychology, 35*, 605–12

7 Sadeh, A., Raviv, A., & Gruber, R. (2000). 'Sleep Patterns and Sleep Disruptions in School-Age Children'. *Developmental Psychology, 36*, 291–301.

8 Wolfson, A. R., & Carskadon, M. A. (2003). 'Understanding Adolescents' Sleep Patterns and School Performance: A Critical Appraisal'. *Sleep Medicine Reviews, 7*, 491–506.

9 http://www.telegraph.co.uk/education/educationnews/7492982/Extra-hour-in-bed-boosts-pupils-exam-results.html.

10 Kuriyama, K., Soshi, T., & Kim, Y. (2010). 'Sleep Deprivation Facilitates Extinction of Implicit Fear Generalization and Physiological Response to Fear'. *Biological Psychiatry, 68*, 991–8.

11 See, for example, Walker, M. P., Brakefield, T., Morgan, A., et al. (2002). 'Practice with Sleep Makes Perfect: Sleep-Dependent Motor Skill Learning'. *Neuron, 35*, 205–11.

12 Gómez, R. L., Bootzin, R., & Nadel, L. (2006). 'Naps Promote Abstraction in Language-Learning Infants'. *Psychological Science, 17*, 670–4.

13 Walker, M. P., Brakefield, T., Morgan, A., et al. (2002). 'Practice with Sleep Makes Perfect: Sleep-Dependent Motor Skill Learning'. *Neuron, 35*, 205–11.

14 Walker, M.P., Brakefield, T., Morgan, A., et al. (2002). 'Practice with Sleep Makes Perfect: Sleep-Dependent Motor Skill Learning'. *Neuron, 35*, 205–11.

15 Stickgold, R., & Walker, M. P. (2007). 'Sleep-Dependent Memory Consolidation and Reconsolidation'. *Sleep Medicine, 8*, 331–43.

16 Walker, M. P., Brakefield, T., Hobson, J. A., Stickgold, R. (2003). 'Dissociable Stages of Human Memory Consolidation and Reconsolidation'. *Nature, 425*, 616–20.

17 Korman, M., Doyon, J., Doljansky, J., Carrier, J., Dagan, Y., et al.

(2007). 'Daytime Sleep Condenses the Time Course of Motor Memory Consolidation'. *Nature Neuroscience*, 10, 1206–13.

18 http://www.sciencedaily.com/releases/2008/06/080609071106.htm.

19 http://www.eurekalert.org/pub_releases/2009-06/aaos-sss060209.php.

20 Mah, C. D., Mah, K. E., Kezirian, E. J., & Dement, W. C. (2011). 'The Effects of Sleep Extension on the Athletic Performance of Collegiate Basketball Players'. *Sleep*, 34, 943–50.

21 Mander, B. A., Rao, V., Lu, B., Saletin, J. M., & Lindquist, J. R. (2013). 'Prefrontal Atrophy, Disrupted NREM Slow Waves and Impaired Hippocampal-Dependent Memory in Aging'. *Nature Neuroscience*, 16, 357.

22 Parent, A. (2004). 'Giovanni Aldini: From Animal Electricity to Human Brain Stimulation'. *The Canadian Journal of Neurological Sciences*, 31, 576–84.

23 Marshall, L., Mölle, M., Hallschmid, M., & Born, J. (2004). 'Transcranial Direct Current Stimulation During Sleep Improves Declarative Memory'. *Journal of Neuroscience*, 24, 9985–92.

24 Ukraintseva, IuV, & Dorokhov, V. B. (2011). 'Effect of Daytime Nap on Consolidation of Declarative Memory in Humans'. *Zh Vyssh Nerv Deiat Im I P Pavlova*, 61, 161–9.

25 Fischer, S., Hallschmid, M., Elsner, A. L., et al. (2002). 'Sleep Forms Memory for Finger Skills'. *Proceedings of the National Academy of Sciences*, 99, 11987–91.

26 Lahl, O., Wispel, C., Willigens, B., & Pietrowsky, R. (2008). 'An Ultra Short Episode of Sleep is Sufficient to Promote Declarative Memory Performance'. *Journal of Sleep Research*, 17, 3–10.

27 Kurdziel, L., Duclos, K., & Spencer, R. M. C. (2013). 'Sleep Spindles in Midday Naps Enhance Learning in Preschool Children'. *Proceedings of the National Academy of Sciences*, doi: 10.1073/pnas.1306418110

28 Rosekind, M. R., Smith, R. M., Miller, D. L., et al. (1995). 'Alertness Management: Strategic Naps in Operational Settings'. *Journal of Sleep Research*, 4, 62–6.

29 Milner, C., & Cote, K. (2009). 'Benefits of Napping in Healthy Adults: Impact of Nap Length, Time of Day, Age, and Experience with Napping'. *Journal of Sleep Research*, 18(2), 272–81.

30 Naska, A., Oikonomou, E., Trichopoulou, A., Psaltopoulou, T., &

Trichopoulos, D. (2007). 'Siesta in Healthy Adults and Coronary Mortality in the General Population'. *Archives of Internal Medicine*, 167, 296–301.

31 Zaregarizi, M., Edwards, B., George, K., et al. (2007). 'Acute Changes in Cardiovascular Function During the Onset Period of Daytime Sleep: Comparison to Lying Awake and Standing'. *Journal of Applied Physiology*, 103, 1332–8.

32 Zhao, D., Zhang, Q., Fu, M., Tang, Y., & Zhao, Y. (2010). 'Effects of Physical Positions on Sleep Architectures and Post-Nap Functions Among Habitual Nappers'. *Biological Psychology*, 83(3), 207–13.

33 Brooks, A., & Lack, L. (2006). 'A Brief Afternoon Nap Following Nocturnal Sleep Restriction: Which Nap Duration is Most Recuperative?'. *Sleep*, 29, 831–40.

34 Bayer, L., Constantinescu, I., Perrig, S., Vienne, J., Vidal, P., Mühlethaler, M., & Schwartz, S. (2011). 'Rocking Synchronizes Brain Waves During a Short Nap'. *Current Biology*, 21(12), R461–R462.

35 Rasch, B., Büchel, C., Gais, S., & Born, J. (2007). 'Odor Cues During Slow-Wave Sleep Prompt Declarative Memory Consolidation'. *Science*, 315, 1426–9.

36 Antony, J. W., Gobel, E. W., O'Hare, J. K., Reber, P. J., & Paller, K. A. (2012). 'Cued Memory Reactivation During Sleep Influences Skill Learning'. *Nature Neuroscience*, 15, 1114–16.

Lesson 6: Welcome to dreamland

1 Cartwright, R. D. (2010). *The Twenty-four Hour Mind: The Role of Sleep and Dreaming in Our Emotional Lives*. Oxford: Oxford University Press.

2 Morewedge, C. K, & Norton, M. I. (2009). 'When Dreaming is Believing: The (Motivated) Interpretation of Dreams'. *Journal of Personality and Social Psychology*, 96, 249–63

3 Much of the information in the section is from: Ullman, M., Krippner, S., & Vaughan, A. (1989). *Dream Telepathy: Experiments in Nocturnal ESP*, 2nd edn., Jefferson, NC: McFarland.

4 Krippner, S., Honorton, C., & Ullman, M. (1973). 'An Experiment in Dream Telepathy with "The Grateful Dead"'. *Journal of the American Society of Psychosomatic Dentistry and Medicine*, 20, 9–17.

5 This procedure is based on a technique described in: Graff, D. E. (2007). 'Explorations in Precognitive Dreaming'. *Journal of Scientific Exploration*, 21, 707–22.

6 Wiseman, R., West, D., & Stemman, R. (1996). 'An Experimental Test of Psychic Detection'. *Journal of the Society for Psychical Research*, 61(842), 34–45.

7 Adler, J. (2006). 'Freud is Not Dead'. *Newsweek*, 147(27 March), 42–6.

8 Schredl, M. (2000). 'Use of Popular Dream Literature'. *Perceptual and Motor Skills*, 90, 1194.

9 Mazzoni, G. A. L., Lombardo, P., Malvagia, S., & Loftus, E. F. (1999). 'Dream Interpretation and False Beliefs'. *Professional Psychology: Research and Practice*, 30, 45–50.

10 Domhoff, G. W. (2000). 'Moving Dream Theory Beyond Freud and Jung'. Paper presented to the symposium Beyond Freud and Jung?, Graduate Theological Union, Berkeley, CA, 23 September 2000.

Lesson 7: Dream therapy

1 Much of the information in this section is from: Domhoff, G. W. (1996). *Finding Meaning in Dreams: A quantitative approach*. New York and London: Plenum Press.

2 Cartwright, R. D., Lloyd, S., Knight, S., & Trenholme, I. (1984). 'Broken Dreams: A Study of the Effects of Divorce and Depression on Dream Content'. *Psychiatry*, 47, 251–259; and Starker, S., & Jolin, A. (1982). 'Imagery and Fantasy in Vietnam Veteran Psychiatric Inpatients'. *Imagination, Cognition and Personality*, 2, 15–22.

3 Bulkeley, K., & Kahan, T. L. (2008). 'The Impact of September 11 on Dreaming'. *Consciousness and Cognition: An International Journal*, 17, 1248–56.

4 Schredl, M. (2011). 'Toilet dreams: Incorporation of waking-life memories?' *International Journal of Dream Research*, 4, 41–4.

5 Yu, C. K.-C., & Fu, W. (2011). 'Sex Dreams, Wet Dreams, and Nocturnal Emissions'. *Dreaming*, 21(3), 197–212.

6 Van De Castle, R. L. (1994). *Our Dreaming Mind*. New York: Ballantine Books.

7 Greenberg, R., Pillard, R., & Pearlman, C. (1972). 'The Effect of

Dream (Stage REM) Deprivation on Adaptation to Stress'. *Psychosomatic Medicine,* 34, 257–262.

8 Cartwright, R., Luten, A., Young, M., Mercer, P., & Bears, M. (1998). 'The Role of REM Sleep and Dream Affect in Overnight Mood Regulation: A Study of Normal Volunteers'. *Psychiatry Research,* 81, 1–8.

9 Cartwright, R. (1996). 'Dreams and Adaptation to Divorce'. In Barrett, D., ed. *Trauma and dreams.* Cambridge, MA: Harvard University Press, 179–185.

10 See, for example, Kupfer D. J. (1976). 'REM Latency: A Psychobiologic Marker for Primary Depressive Disease'. *Biological Psychiatry,* 11, 159–74.

11 This is not the only theory that has been proposed to account for why those diagnosed with depression spend longer in REM, and why REM-deprivation helps alleviate their symptoms. For a comprehensive discussion about the topic, see Cartwright, R. D. (2010). *The Twenty-four Hour Mind: The Role of Sleep and Dreaming in Our Emotional Lives.* Oxford: Oxford University Press.

12 Vogel, G. W., Thurmond, A., & Gibbons, P. (1975). 'REM Sleep Reduction Effects on Depression Syndromes'. *Archives of General Psychiatry,* 32, 765–77.

13 Wilson, S., & Argyropoulos, S. (2005). 'Antidepressants and Sleep: A Qualitative Review of the Literature'. *Drugs,* 65(7), 927–47; and Mayers, A. G., & Baldwin, D. S. (2005). 'Antidepressants and Their Effect on Sleep'. *Human Psychopharmacology,* 20(8), 533–59.

14 Dement, W. C., & Vaughan, C. (1999). *The Promise Of Sleep: A pioneer in sleep medicine explores the vital connection between health, happiness, and a good night's sleep.* New York: Delacorte Press.

15 Hill, C. E., & Rochlen, A. B. (2004). 'The Hill Cognitive-Experiential Model of Dream Interpretation'. In Rosner, R., Lyddon, W, & Freeman, A., eds. *Cognitive Therapy and Dreams.* New York: Springer Books, 75–89.

16 Hill, C. E., & Goates, M. K. (2004). 'Research on the Hill Cognitive-Experiential Dream Model'. In Hill, C. E., ed. *Dream Work in Therapy: Facilitating Exploration, Insight, and Action.* Washington, DC: American Psychological Association, 245–288.

17 Hill, C. E., Zack, J., Wonnell, T., Hoffman, M. A., Rochlen, A.,

Goldberg, J., Nakayama, E., Heaton, K. J., Kelley, F., Eiche, K., Tomlinson, M., & Hess, S. (2000). 'Structured Brief Therapy with a Focus on Dreams or Loss for Clients with Troubling Dreams and Recent Losses'. *Journal of Counseling Psychology*, 47, 90–101.

18 Kolchakian, M. R., & Hill, C. E. (2002). 'Dream Interpretation with Heterosexual Dating Couples'. *Dreaming*, 12, 1–16.

19 Falk, D. R., & Hill, C.E. (1995). 'The Effectiveness of Dream Interpretation Groups for Women Undergoing a Divorce Transition'. *Dreaming*, 5, 29–42.

20 Zanasi, M., DeCicco, T. L., Murkar, A., Longo, G., & Testoni, F. (2010). 'Waking Day Anxiety and Dreams: Dream Content and Predictors in Italian and Canadian Samples'. *International Journal of Dream Research*, 3, 12–13; and DeCicco, T. L., & Higgins, H. (2009). 'The Dreams of Recovering Alcoholics: Mood, Dream Content, Discovery, and The Storytelling Method of Dream Interpretation'. *International Journal of Dream Research*, 2, 45–51.

21 Dement, W. C. (1972). *Some Must Watch While Some Must Sleep*. New York: Norton.

22 White, G. L., & Taytroe, L. (2003). 'Personal Problem-Solving Using Dream Incubation: Dreaming, Relaxation, or Waking Cognition?'. *Dreaming*, 13, 193–209.

23 Cai, D. J., Mednick, S. A., Harrison, E. M., Kanady, J. C., & Mednick, S. C. (2009). 'REM, Not Incubation, Improves Creativity by Priming Associative Networks'. *Proceedings of the National Academy of Sciences*, 106, 10130–4.

24 Cai, D. J., Mednick, S. A., Harrison, E. M., Kanady, J. C., & Mednick, S. C. (2009). 'REM, Not Incubation, Improves Creativity by Priming Associative Networks'. *Proceedings of the National Academy of Sciences*, 106, 10130–4.

25 Barrett, D. (1993). 'The "Committee Of Sleep": A Study Of Dream Incubation For Problem Solving'. *Dreaming*, 3, 115–22.

26 Stevenson, R. L. (1924). *The Works of Robert Louis Stevenson*, Tusitala Edition, Vol 30. London: Heinemann, 41–53.

27 Scott, W. (1890/1972). *The Journal of Sir Walter Scott, Bart.*, Anderson, W. E. K., ed. Oxford: Clarendon Press.

28 Schock, J., Cortese, M. J., & Khanna, M. M. (2012). 'Imageability

estimates for 3,000 disyllabic words'. *Behavior Research Methods*, 44(2), 374–9.

29 Baine, D. (1986). *Memory and instruction*. Englewood Cliffs, NJ: Educational Technology Publications.

30 http://www.mheap.com/gibson.html.

31 Gibson, H. B. (1985). 'Dreaming and Hypnotic Susceptibility: A Pilot Study'. *Perceptual and Motor Skills*, 60, 387–94; and Zamore, N., & Barrett, D. (1989). 'Hypnotic Susceptibility and Dream Characteristics'. *Psychiatric Journal of the University of Ottawa*, 14, 572–4.

32 Spiegel, H. (1974). 'The Grade 5 Syndrome: The Highly Hypnotizable Person'. *International Journal of Clinical and Experimental Hypnosis*, 22, 303–19; and Spiegel, H., & Spiegel, D. (1978). *Trance & Treatment: Clinical uses of hypnosis*. New York: Basic Books.

Lesson 8: Sweet Dreams

1 LaBerge, S. P. (1980). 'Lucid Dreaming: An Exploratory Study of Consciousness During Sleep'. Unpublished doctoral dissertation. Stanford University, US; and LaBerge, S. P., Nagel, L. E., Dement, W. C., & Zarcone, V. P. J. (1981). 'Lucid Dreaming Verified by Volitional Communication During REM Sleep'. *Perceptual and Motor Skills*, 52, 727–32.

2 Stumbrys, T., Erlacher, D., Schädlich, M., & Schredl, M. (2012). 'Induction of Lucid Dreams: A Systematic Review of Evidence'. *Consciousness and Cognition*, 21, 1456–75.

3 Gackenbach, J. (2006). 'Video Game Play and Lucid Dreams: Implications for the Development of Consciousness'. *Dreaming*, 16, 96–110.

4 Schredl, M., & Erlacher, D. (2011). 'Frequency of Lucid Dreaming in a Representative German Sample'. *Perceptual and Motor Skills*, 112, 104–8.

5 Tholey, P. (1989). 'Consciousness and Abilities of Dream Characters Observed During Lucid Dreaming'. *Perceptual and Motor Skills*, 68, 567–78.

6 Stumbrys, T., Erlacher, D., & Schmidt, S. (2011). 'Lucid Dream

Mathematics: An Explorative Online Study of Arithmetic Abilities Of Dream Characters'. *International Journal of Dream Research*, 4, 35–40.

7 Erlacher, D., & Schredl, M. (2010). 'Practicing a Motor Task in a Lucid Dream Enhances Subsequent Performance: A Pilot Study'. *The Sport Psychologist*, 24, 157–67.

8 Agargun, M. Y., Boysan, M., & Hanoglu, L. (2004). 'Sleeping Position, Dream Emotions, and Subjective Sleep Quality'. *An International Journal of Sleep, Dream, and Hypnosis*, 6, 8–13.

9 Van de Castle, R. L. (1994). *Our Dreaming Mind*. New York: Ballantine Books, 210.

10 Schredl, M., et al. (2009). 'Information Processing During Sleep: The Effect of Olfactory Stimuli on Dream Content and Dream Emotions'. *Journal of Sleep Research*, 18, 285–90.

11 Castaldo, V., & Holzman, P. (1967). 'The Effects of Hearing One's Own Voice on Sleep Mentation'. *Journal of Nervous and Mental Disease*, 144, 2–13; and Castaldo, V., & Holzman, P. (1968). 'The Effects of One's Own Voice on Dream Content'. *Psychophysiology*, 5, 218–19.

12 Cajochen, C., Altanay-Ekici, S., Münch, M., Frey, S., Knoblauch, V., & Wirz-Justice, A. (2013). 'Evidence that the Lunar Cycle Influences Human Sleep'. *Current Biology*, 23(15), 1–4.

13 Wegner, D. M., & Schneider, D. J. (2003). 'The White Bear Story'. *Psychological Inquiry*, 14, 326–29.

14 Harvey, A. G. (2003). 'The Attempted Suppression of Presleep Cognitive Activity in Insomnia'. *Cognitive Therapy and Research*, 27, 593–602.

15 Bryant R. A., Wyzenbeek, M., & Weinstein, J. (2011). 'Dream Rebound of Suppressed Emotional Thoughts: The Influence of Cognitive Load'. *Consciousness and Cognition*, 20(3), 515–22.

16 Krakow, B., Kellner, R., Pathak, D., & Lambert, L. (1995). 'Imagery Rehearsal Treatment for Chronic Nightmares'. *Behaviour Research and Therapy*, 33, 837–43; Neidhardt, E. J., Krakow, B., Kellner, R., & Pathak, D. (1992). 'The Beneficial Effects of One Treatment Session and Recording of Nightmares on Chronic Nightmare Sufferers'. *Sleep*, 15, 470–3; and Kellner, R., Neidhardt, J., Krakow, B., & Pathak, D. (1992). 'Changes in Chronic Nightmares after One Session of Desensitization or Rehearsal Instructions'. *American Journal of Psychiatry*, 149, 659–63.

17 Krakow, B. J., Melendrez, D. C., Johnston, L. G., et al. (2002). 'Sleep Dynamic Therapy for Cerro Grande Fire Evacuees with Posttraumatic Stress Symptoms: A Preliminary Report'. *Journal of Clinical Psychiatry*, 63, 673–84.

18 Krakow, B., Hollifield, M., Johnston, L., et al. (2001). 'Imagery Rehearsal Therapy For Chronic Nightmares in Sexual Assault Survivors with Posttraumatic Stress Disorder: A Randomized Controlled Trial'. *Journal of the American Medical Association*, 286, 537–45.

19 Wiseman, A. S. (1989). *Nightmare Help: A Guide for Parents and Teachers*. Berkeley: Ten Speed Press.

Acknowledgements

First and foremost, I wish to thank the University of Hertfordshire for supporting my work over the years. I would like to thank Clive Jefferies and Emma Greening for reading earlier drafts of the manuscript. My thanks also to Stevie Williams, Chris Idzikowski, Stephen LaBerge, Stanley Krippner, Rosalind Cartwright, and Jenny Hambelton for their invaluable contributions. This book would not have been possible without the expert guidance and expertise of my agent Patrick Walsh and editor Jon Butler. Thanks to DOW for producing the illustrations, and special thanks also to my wonderful colleague, collaborator, and partner, Caroline Watt. Finally, I would like to thank the many sleep scientists who dedicated their lives to exploring the night, and the thousands of volunteers who were kind enough to go along on the ride.

Appendix

Donald and the Elephant at School

Jenny Hambelton and Richard Wiseman

One sunny day Donald was walking happily to school. It was an exciting day because the children were going on a trip to the new park in the afternoon on a big, shiny bus.

'I have a funny feeling that today is going to be my lucky day,' thought Donald.

Bump! Donald walked straight into a large grey elephant.

'Hello there,' said the elephant in a friendly voice. 'I have never been to a school. Can I come to school with you?'

'Yes, of course you can,' said Donald.

When the bell went, Donald took the elephant into class.

'Your elephant can only stay if he is very good and helpful,' said the teacher.

'I am sure he will be,' said Donald.

At painting time the elephant put his apron on and sucked up all the paint into his trunk.

He sprayed it all over the teacher.

'This is a disaster!' thought Donald.

At snack time the children offered the fruit bowl to the elephant to choose a piece of fruit. He picked up all the fruit and ate it in one gulp.

'This is a disaster!' thought Donald.

It was time for swimming. Donald lent the elephant a pair of trunks and the elephant jumped in the pool. When the elephant jumped in, all the water in the pool came out.

'This is a disaster!' thought Donald.

Before lunch, it was time for singing but the elephant trumpeted so loud that all the children put their hands on their ears.

'This is a disaster!' thought Donald.

The teacher was cross. 'I said your elephant could only come to school if he was kind and helpful. So far he has covered me in paint, eaten all the fruit, emptied the swimming pool, and made everyone's ears hurt with his terrible singing. This is a disaster!'

Donald was sad. He'd thought it was going to be a lucky day and everything was going wrong.

'Please give him one more chance,' said Donald. 'He wants to be very kind and helpful.'

At last it was time for the trip! The children climbed onto the bus and the elephant put his seat belt on and sat next to Donald. They had only been going a little way when one of the tyres went 'POP'.

'This is a disaster!' thought Donald.

The elephant got off the bus and blew the tyre up with his trunk.

'I knew he would be very kind and helpful,' said Donald.

When they got to the park the gate was locked.

'Oh no,' said all the children. 'This is a disaster!'

'Don't worry,' said the teacher. 'I have the keys.'

Just as he said this, one of the children bumped into the teacher. He dropped the keys and they fell down the drain.

'Oh no,' said all the children. 'This is a disaster!'

Quick as a flash, the elephant reached down the drain with his long trunk and carefully lifted the keys out and gave them to the teacher. Everyone cheered and agreed that after all it had been very lucky that Donald had met the elephant on his way to school that day.

'It just goes to show,' said the teacher, 'That sometimes things may seem bad but they often turn out for the best!'

On the way home, Donald wondered if the elephant would come back to school the next day. But he never saw him again.

After you have read the story, help cement its meaning in place by saying something like 'Well, Donald thought that taking the elephant to school was a disaster, but it turned out for the best – it just goes to show how bad things often work out well in the end.'